Critical Studies of the Asia-Pacific

Series editor
Mark Beeson
Department of Political Science & International Relations
University of Western Australia
Crawley, WA, Australia

Critical Studies of the Asia Pacific showcases new research and scholarship on what is arguably the most important region in the world in the twenty-first century. The rise of China and the continuing strategic importance of this dynamic economic area to the United States mean that the Asia-Pacific will remain crucially important to policymakers and scholars alike. The unifying theme of the series is a desire to publish the best theoretically-informed, original research on the region. Titles in the series cover the politics, economics and security of the region, as well as focusing on its institutional processes, individual countries, issues and leaders.

More information about this series at
http://www.palgrave.com/gp/series/14940

Marie-Hélène Schwoob

Food Security and the Modernisation Pathway in China

Towards Sustainable Agriculture

Marie-Hélène Schwoob
IDDRI
Paris, France

Critical Studies of the Asia-Pacific
ISBN 978-3-319-65701-1   ISBN 978-3-319-65702-8   (eBook)
https://doi.org/10.1007/978-3-319-65702-8

Library of Congress Control Number: 2017954310

Cover illustration: National Geographic Creative / Alamy Stock Photo

Printed on acid-free paper

This Palgrave Macmillan imprint is published by Springer Nature
The registered company is Springer International Publishing AG
The registered company address is: Gewerbestrasse 11, 6330 Cham, Switzerland

ACKNOWLEDGMENTS

None of this work could have been possible without the precious help of people I would like to express my profound gratitude to.

My first thanks go to my financial partners, without whom I would never have been able to embark upon the arduous path of PhD candidates. The Region Ile-de-France was among the first to see potential in my work and supported me unconditionally during the entire three years of my thesis through its Allocation Doctorale Hors-DIM. The French Ministry of Foreign Affairs (Program Hubert Curien Cai Yuanpei 2011–2015 and 2014 Zhang Heng program) and Asia Centre—I would like to deeply thank Jean-François Di Meglio, in particular, for his constant encouragement—helped me a lot as well by funding my regular fieldtrips to China, without which I would never have been able to gather so much research material.

Landing in China is just the first step of one's research and not much would have come out of my stays without the support of a number of people I had the chance to meet over there. Conducting research in China is not an easy task. All the obstacles one has to overcome would be disheartening without the incredible cheering-up effect of meeting people who are willing to help, because they care, because they share one's concerns, or just out of kindness and openness of mind. In particular, my deep thanks go to Carole Ly, Alain Bonjean, Sylvie Dideron, Gabrielle Harris, Yang Jun, Qiu Huanguang, Zheng Huan, Wu Zhifen, Alex Bulcourt, Michel Humbert, Xavier, April, Clotilde, Fanny, and all the others for their kind help and support. I would also like to sincerely thank Professor Wu Yongping and the School of Public Policy and Management of

Tsinghua University for the opportunity they gave me to come to China as part of the Science Po-Tsinghua Initiative for Public Policy, and to exchange on advanced policy research topics with excellent scholars and PhD candidates who challenged and enriched my ideas.

My gratitude, of course, also goes to my research director, who dedicated a lot of his valuable time to the supervision of my work. Richard Balme knew how to make himself available and gave me countless feedbacks and suggestions that always enlightened me when I did not know where to go anymore. Not only were our discussions incredibly valuable for my research but they were also very pleasant, something for which I am much grateful.

I had the chance to see my work supervised by another brilliant mind. Sebastien Treyer, Director of Programs at Iddri, has informed and shaped my research questions by linking them to the global debates on food security and agricultural development. I could never thank him enough for having shared his knowledge and for having given me the opportunity to work at Iddri, where I learned an incredible amount, enjoyed a remarkable work environment, and met an amazing community of researchers with whom I developed wonderful friendships.

How to talk about friendship without mentioning the ones I made with my fellow PhD candidates and the other sinophiles? Even the 200 pages of this book would not be enough to describe the beauty and the strength of all these shared moments of joy and sorrow, all these laughs, and all these worries. In China and Europe, they made us closer and shaped our personalities, leaving us with incredible memories that will be called up in our minds for years to come.

CONTENTS

ACRONYMS

AQSIQ   Administration of Quality Supervision, Inspection and Quarantine
CABTS   China Agricultural Broadcasting and Television School
CAU   Chinese Agricultural University
CCP   Chinese Communist Party
COFCO   China National Cereals, Oils and Foodstuffs Corporation
CSA   Community-Supported Agriculture
DP   Direct Purchase
FAO   Food and Agriculture Organization of the United Nations
ID (card)   Identification (card)
KMT   Kuomintang
MNE   Multi-National Enterprises
MOA   Ministry of Agriculture
NATESC   National Agricultural Technology Extension and Service Center
NDRC   National Development and Reform Commission
NGO   Non-governmental Organizations
NPK   Nitrogen Phosphor Potash
PRC   People's Republic of China
RMB   Renminbi
SOE   State-Owned Enterprise
TVE   Township and Village Enterprise
VAT   Value-Added Tax

LIST OF FIGURES

LIST OF TABLES

CHAPTER 1

Introduction

China, a developing economy and a major food importing and exporting country, provides us with an extremely interesting example of the complexity and the rising challenges of agricultural modernization. The country, which has to feed almost 20 percent of the world population with only 7 percent of the world arable land, needs sufficient amounts of agricultural commodities at a tolerable price, as the share of food is still high in total consumers spending.[1] Meanwhile, the growing urban middle class is asking for new types of food diet. The resulting stimulation of the national oil and meat consumption has effects on the demand for land intensive products, such as feed and oilseeds. Since the country became a net importer of food in 2004, its agricultural balance has become heavier every day.

Considering the demographic weight of China, the stakes go well beyond the Chinese territory. The growing food insecurity of the country could have disastrous consequences on global food markets and, in the end, on other importing countries. The risks are also substantial for China. Despite the fact that its massive trade surplus theoretically balances rising food imports, relying on global markets for food would put the country's population at greater risk in terms of price volatility. As a consequence, tackling issues related to food security has turned into a real priority for the Chinese government.

The government, urged to implement effective agricultural development and food security policies, has reshaped its political agenda since the

© The Author(s) 2018
M.-H. Schwoob, *Food Security and the Modernisation Pathway in China*, Critical Studies of the Asia-Pacific,
https://Doi.org/10.1007/978-3-319-65702-8_1

beginning of the 2000s, putting agricultural development and food secu-rity back to the forefront of its political priorities. This represents a major shift away urban and industrial development, which was the most impor-tant focus of the last decades of the twentieth century, as a major source of growth both in urban and rural areas (Oi 1999; Lin et al. 2000).

But how to frame agricultural development and food security policies in the twenty-first century? Productivist agricultural practices that have prevailed over the past decades already started deteriorating arable land and water in China, putting even more pressure on already scarce but essential resources for the sustainability of agricultural production on the middle- and long-term. Is the awareness of the government of these issues likely to trigger a policy response and to make agricultural practices evolve toward more sustainable farming practices in China—using less water, less pesticides, and chemical fertilizers and offering better working conditions to farmers? Is an alternative pathway, environmentally and socially more sustainable, likely to become reality in the near future? These are some of the questions this book wishes to address.

1.1   AGRICULTURAL TRANSITIONS, IN THE PAST AND TODAY

What are agricultural transitions? Characterizing change has always been a challenging task, as change encompasses a wide variety of political, social, and economic dimensions. When it comes to agricultural change, a large body of literature exists that helps to better assess it. The different bodies of literature on agricultural transition do not necessarily refer to the same notions of transition. At least four different corpuses can be distinguished, which depict different transition processes, sometimes overlapping. A first corpus focuses on agricultural transitions in socialist and communist econ-omies evolving toward market economies (Swinnen and Rozelle 2006). A second one focuses on agricultural transitions in developing countries, where agricultural modernization is usually depicted as the first step of an economic development path entitled the "Lewis" path—although this has recently been put back into question (Dorin et al. 2013), as we will see later in this book. This second corpus developed a lot during the spread of the Green Revolution in the 1960s, 1970s, and 1980s—an agricultural development paradigm aimed at increasing yields through the use of high-yielding varieties, chemical fertilizers and agrochemicals such as pesticides and herbicides, and sometimes through irrigation and mechanization. A

Table 1.1   Different definitions for "agricultural transition"

	Definition 1	*Definition 2*	*Definition 3*	*Definition 4*
Country/ economy	Socialist/ communist economies	Developing countries	Developing countries	Developed and developing economies
Objective	Transition to market economy	Agricultural modernization as a first step of Lewis-type economic development	Integration in international markets	Transition toward more sustainable and more productive models

third body of literature—sometimes associated with the one on agricultural modernization in developing countries—concentrates on agricultural transitions in a given country (usually a developing economy) willing to integrate international markets (usually for World Trade Organization [WTO] integration purposes). Finally, a last body of literature focuses on agricultural transition toward more productive *and more sustainable* models (Table 1.1).

The body of literature on agricultural transitions toward more sustainable and more productive models has developed a lot over the past few years, especially since the 2007–2008 food price crisis. As a consequence of soaring food prices—in 2008, the cereal price index reached a peak 2.8 times higher than in 2000 (United Nations 2011: 62)—an estimated 44 million people were driven into poverty (World Bank 2011), and many countries were confronted to major social and political crises. Six years after the food price crisis, agricultural issues are still to be addressed, both in developing and in developed countries. The question of how to provide food, at a decent price, to 9 billion people by 2050, is a matter of intense debates and an important number of people and organizations have been urging countries to raise their agricultural productivity levels. However, in a context where arable land and water resources are limited and already eroded by the rising needs of urbanization and industrialization, agricultural intensification has turned into an additional threat to the sustainable use of these resources. As a consequence, a growing number of people and organizations advocate in favor of a transition toward more sustainable agriculture. The debate seems to have polarized around two extremes: the advocates of productivism, for whom the main goal of agricultural policies should be to raise production levels in order to feed the ever-increasing world population, and the proponents of environmental protection, for whom the implementation of sustainable farming practices should be considered as a priority to lower the impacts of agricultural production on the environment. In reality, the array of movements is much larger than these sole two poles. Holt-Gimenez and Shattuck (2011), for instance, acknowledge at least four main categories of opposing "global food movements": the "neoliberal" one, the "reformist" one, the "progressive" one, and the "radical" one. While the neoliberal movement is based upon a discourse oriented toward corporate and global markets and giving priority to "food enterprises", the reformist movement, on its side, gives priority to food security, development, and aid. The progressive movement, primarily based in northern countries, relies on a "food justice discourse" that promotes the development of local foodsheds, family farming, and access to fresh and affordable food, with a strong emphasis on direct rural–urban linkages and alternative business models that insist on social rather than individual (consumers') responses to food regime failures. Finally, the radical movement, which endorses some of the elements of the progressive movement, advocates in favor of deep and structural changes of agriculture and food systems toward more sustainability, more fairness, more sovereignty, and more security.

In China, the debate between productivism and sustainability is vivid. The government, which long had to deal with insufficient resources, clearly keeps on attaching fundamental importance to the capacity of the territory to supply the demand of the population. On the other side, rural industrialization and intensive agriculture had dramatic consequences on soil, water and on the safety of food products. Environmental protection recently emerged as a strongly debated element for the pathway which the agricultural sector is embarking on. Internationally, debates on the new pathways of agricultural development and transition have intensified, especially since the food crisis of 2007–2008. The question of which path should agricultural modernization take is thus not unique to China. Exploring the modalities of agricultural transition pathways in China provides substantial elements for the understanding of the building of national agricultural pathways worldwide.

1.2   A NEED FOR SOCIO-POLITICAL APPROACHES

While the productivist paradigm of the Green Revolution was mostly based on the spread of technology aimed at improving yield and farmers' income, the current paradigm of "sustainable agriculture" is way more complex and diverse. The question of the adoption of more sustainable farming practices is not about addressing short-term economic issues anymore, and cruelly needs sociological answers. What makes farmers adopt more sustainable practices, of which the benefits are sometimes only seen on the long term? Which lock-ins prevent the transformation of agriculture toward more sustainable systems?

A solid literature is currently developing on the subject, particularly on the lock-ins created by agricultural "technological pathways". David (1985) demonstrated that irreversibilities due to technical interrelatedness, scale economies, and learning and habituation could be brought by the adoption and standardization of a technological system. Similarly, a number of agronomists have shown that escaping from agricultural systems based on technology such as chemical inputs was a slow process (Barbier and Elzen 2012; Vanloqueren and Baret 2008). However, in spite of an extremely rich and broad literature on path dependencies in political science and sociology, there is little dialogue between the two corpuses. This book aims at filling this gap, by relying on the theoretical frameworks of policy and sociological analysis to explore China's transition toward more sustainable agricultural systems.

As the country already conducted its green revolution—basically meaning that farmers possess the technical means (such as pesticides and fertilizers) to improve productivity and already use them extensively—levers of action to increase agricultural production are now essentially to be found in agricultural structures and practices (see Table 1.2). The Chinese farming structure is indeed still characterized by small-scale agriculture[2] poorly suited for mechanized agriculture and economies of scale. Therefore, "reorganizing" stakeholders taking part in agricultural production has become a necessity to carry out China's "new agricultural modernization".

This "sociological side" (the "reorganization" of producers), which recent agricultural modernization policies tend to focus on, is likely to have a strong impact on patterns of relationships in rural areas. This research topic attracted my attention as a still relatively unexplored issue. Public policies never apply on a "neutral" substrate. Analyzing patterns of power in rural areas is therefore key to shed light on the modalities of the change occurring in the course of agricultural modernization.

Change does not only act *upon* actors. A large body of literature evidences the *active* role social actors play in institutional change—and in lock-ins as well. According to Bezès and Le Lidec (2010: 58), the emergence of institutional reforms is facilitated by social actors they call "reform entrepreneurs",

Table 1.2 Levers of actions to increase agricultural production in China

Production factor	Possibility to act as a lever	Main obstacles preventing the possibility to act as a lever
Arable land (quantity)	No	Urbanization, desertification
Arable land (quality)	No	Pollution, desertification, unsustainable agricultural practices, and over-exploitation
Pesticides, fertilizers (quantity/ha)	No	Current situation of over-consumption
Pesticides, fertilizers (spreading techniques)	Weak	Lack of vocational training, imperfections of extension services, highly subsidized industries
Irrigation	Weak-strong	Desertification, lack of investment capacities for local small irrigation and water-saving infrastructures
Mechanization	Strong	Lack of investment capacities for small farmers, mountainous, and hilly areas
Organization, cooperation and economic rationalization	Strong	Social, institutional, and political roadblocks
Science and technology (GMO and hybrid varieties)	Strong	Intellectual property issues, civil society concerns, investment barriers for small farmers, question marks for the sustainability of the model

who are in a position to transform institutional rules by demonstrating their ability to provide answers to address a given issue and by building support coalitions. Similarly, Sabatier and Jenkins-Smith (1993) developed the Advocacy Coalition Framework to analyze the "social" causes of the emergence of policy change on long-term time frames. According to the authors, actors who share basic ontological and normative beliefs are grouped in coalitions, within which they develop strategies to transform their beliefs in concrete public policies. Another major framework used to depict policy change from a sociological point of view is the Epistemic Community Framework. Developed by Haas (1992), it depicts how networks of knowledge-based experts help governments identify their interests and frame the collective debates, considerably influencing policy-making. For Stone Sweet, Fligstein and Sandholtz (2001: 11), "skilled actors", who "find ways to induce cooperation amongst disparate individuals or groups by helping them to form a stable conception of roles and identity" are among the four main causes of institutional change.

The importance of the role played by social actors in the course of institutional change goes well beyond the stage of the emergence of a reform.

Social actors also play an important part in the implementation of change. According to Mayntz (1980), three different dimensions determine the effectiveness of policy (or policy reform) implementation: the choices made in the program design concerning intervention instruments, the procedural and organizational arrangements of the administrative implementation structure, and the situation and evolution of the social environment (e.g., the economic, political, and social weight of the groups targeted by the new policy). In the 1980s, an important body of literature developed on this last dimension. This new approach, called "bottom-up" by Knoepfel, Larrue, and Varone (2001: 222), was first initiated by Hjern and Hull (1982) and would be opposed to a "top-down approach" led by Sabatier and Mazmanian (1979). This "bottom-up" approach advocates for a mainstreaming of the analysis of interactions between social actors (target groups, third party groups, and other players) as the first step of research on the effectiveness of policy implementation. The role played by stakeholders in reform processes and change is thus supported by a large corpus of literature, reinforcing the legitimacy of the methodology chosen for this research, which places great emphasis on sociological analysis.

The role played by the state and public policies in agriculture is no less important. Agricultural policies need to answer the national stake of food security—a stake, which ruling regimes could never choose to ignore without risking their collapse. This is particularly true for China, as the country's history, scarred by ancient and recent famines, deeply engraved a fear of food shortage in the minds of central government officials. Although in the current context of globalized markets, famines are not the main fear of most governments anymore, agriculture is still a matter of public affairs. In fact, the return of Western states in the agricultural sector can be traced back to the middle of the twentieth century. For Muller (1984), this comeback is explained by the emergence of a global frame of reference for agricultural modernization in favor of productivism in the 1950s and 1960s. This productivist movement not only created space for the action of new associative structures and private stakeholders but also made the involvement of the state in agriculture stronger and more likely to last on the long term. Public subsidies, established to support the trade of agricultural products after the crisis of the 1930s (Mollard 2002), are still substantial today. The massive share of the budget of the European Union dedicated to agriculture (38% for the CAP 2014–2020) and the annual spending of the US government for the Farm Bill policy are clear proofs that central governments are still strongly involved in agricultural affairs.

1.3   Research Design

In what ways do the interactions between state and nonstate actors triggered by the restoration of the state's involvement in agricultural production activities frame the agricultural modernization pathway which China is engaging on?

The state cannot be considered as an entity that can be looked at in isolation from nonstate actors. According to Migdal's (2001: 12) state-in-society perspective, "states are no different from any other formal organizations or informal social grouping". In agreement with Midgal, this research recognizes that the state is not a "coherent, integrated, and goal-oriented body" and that the state and society mutually transform each other and build from one another. As Remick (2004: 5) puts it, "the state is purely an organization", and this research intends to analyze it as such. "Images of the state" were put aside as much as possible, in order to focus on concrete practices perpetuated by state actors: government officials of township and county levels, tied to the party to a greater or lesser extent. This deconstruction of the Chinese state was particularly useful to reach conclusions on practices enabling governmental actors to reinvestigate the agricultural sector but could not go without asking questions about what, in the end, was holding the multiple actors of the Chinese state together. Considering the state, at the same time, as a social relation influenced by patterns of relationships and as a working entity shaping its environment, is a dichotomy that this research builds on.

The frames of reference promoted by agricultural modernization policies and the social patterns these frames are plugged onto constitute the two main objects of analysis of this research. In order to approach these two objects, it was necessary to conduct two kinds of fieldwork. The first one concentrated on the modalities of implementation of agricultural modernization in rural areas. The second one focused on the definition of the frames of reference of agricultural modernization at the central level of the government. In total, close to 200 stakeholders were interviewed for the purpose of this research.

The first fieldwork explored the various realizations of interactions between stakeholders taking part in the modernization of agricultural production at the local level. Although agricultural production is still mostly taken care of by small farmers, agricultural enterprises play an increasing role in the picture. In the past, much research conducted on rural areas drew conclusions on the interactions between a limited number of stakeholders. "State-peasants" relationships were perhaps the most studied

(Ash 2006; Cai 2000; Bernstein and Lu 2000; see, as well, the corpus of literature depicting the consequences of the abolition of peasants' burden on local governance patterns with Tao et al. 2011; Kennedy 2007, etc.). "State-enterprises" relationships were also rather thoroughly analyzed in rural areas (Unger and Chan 1999). However, farmers and industrial players both take part in agricultural production in rural areas. As a consequence, a more comprehensive analysis of the whole concrete system of action (agricultural production), at the local level, appeared necessary to fully understand the picture of agricultural modernization.

Agricultural modernization constitutes a unique framework where numerous stakeholders meet and interact. The methodology developed by Crozier and Friedberg (1977) rapidly appeared as a suitable frame for analysis. This approach, which was a fundamental contribution to the study of organizations and change, is based on the analysis of a concrete system of action—local agricultural production in this research. The system of action is made of strategic actors, who interact with each other according to the features of the system, to their own interests and to their own resources, which depend on their capacity to control the uncertainties of the system. Information obtained through interviews with strategic actors of the system helps to understand "how each actor confronts his situation and its inherent constraints, *what objectives* he sets for himself, and how he perceives his potential for attaining these objectives within a given structure". In other words, interviews look "what *resources* the actor possesses, what his *margin of liberty* is, and *in what way, under what conditions,* and *within what limits* he can make use of them" (Crozier and Friedberg 1977: 263).

The analysis of the interests, resources, and strategies of stakeholders engaged in local agricultural production, as well as the analysis of the uncertainties of the system and of the capacity of each actor to control these uncertainties, were part of a first step of research aimed at gaining an accurate picture of local patterns of power. Logics of association, partnership, interdependence, and latent conflict progressively appeared along fieldwork analysis. The understanding of the patterns of power and of the relationships between the actors of the concrete system of action of agricultural production shed light on the different ways used by state actors to restore their presence in agricultural production activities, and on how local players were reacting to this new agricultural modernization program.

Actors are rational. They have reasons to behave as they do and deploy strategies according to their interests and to the situation as they perceive it. Interests and preferences are not set in absolute terms. Rather, they vary according to institutional contexts (Hall 1996) and interactions between actors, who behave in an opportunistic manner (Steinmo et al. 1992) with bounded rationality. In order to analyze actors in the selected concrete system of action, particular importance was attached to the analysis of the context-related resources, context-related interests, context-related preferences, and context-related strategies of stakeholders.

In *Capitalism from Below*, Nee and Opper (2012: 69) perfectly illustrate how the rise of entrepreneurship in the Yangzi delta region "was not fueled by exogenous institutional changes", but rather by entrepreneurs themselves who developed and used "innovative informal arrangements within close-knit groups of like-minded actors that provided the necessary funding and reliable business norms that allowed the first wave of entrepreneurs to survive outside the state-owned manufacturing system". This research, as well, by relying on an actor-centered approach, attaches importance to the fact that institutional frameworks shaped by the state do not entirely define the behavior of nonstate actors, and analyzes the strategies deployed by local stakeholders to use these frameworks or act despite or outside them.

The dynamic analysis of the concrete system of action of agricultural production was conducted in several case study areas in the countryside. Counties (*xian* 县), "the strongest and most coherent subprovincial administrative unit" and "the foundation of China's national government" (Blecher and Shue 1996: 204) appeared to be suitably sized areas for this research, as counties' prominent role in rural affairs was underlined by a number of scholars (Lam 2010).

China is one of the largest countries worldwide. With latitudes between 18° and 54° N and an impressive geographic variety, the territory includes an important number of climate types. The diverse range of natural environments and climates enables the country to cultivate a wide array of agricultural products, from pineapples on the tropical island of Hainan to maize, wheat, and grass-fed livestock in the provinces of the North.

The diversity of products comes along with a diversity of farming methods. A number of these latest were depicted in the amazing book written by King (1949: 15), *Farmers of Forty Centuries*. Methods include irrigation systems, selection of crop varieties adapted to local conditions, methods of fertilization, and a wide range of other traditional farming practices, in sum, an "unimpaired inheritance moving with the momentum acquired through 4,000 years" that enabled Asian countries to maintain the fertility of its soil.

Five broad agricultural regions can be depicted:

- The first one includes the mountainous provinces and autonomous region of Inner Mongolia, Gansu, Xinjiang, Qinghai, and Tibet. It is mainly made of grazing areas used by pastoral farmers for meat, milk, wool, and cashmere production, with lowland regions famous for their specialized agriculture, mainly producing cotton, sunflower, rapeseeds, and tomatoes.
- The second agricultural region includes the north-eastern provinces of Heilongjiang, Liaoning and Jilin, which produce mainly grain, such as maize, wheat, sorghum, and soybean. In these areas, only one harvest per year is possible because of the cold and harsh climate.
- The third zone is located on the borders of the Huang and Huai rivers, in the North of China. The area is highly specialized in wheat production (e.g., Henan produces one-third of the wheat produced in the whole country). The regions located at the south of the area can yield two crops per year (usually rice, maize, sorghum, soybean or fruits, and vegetables).
- The fourth agricultural region is located on the borders of the Yangzi River, with rice as the main crop, and grain, fruits, or cash crops such as tea plantation with secondary crops.
- The fifth area is located at the extreme south of the country. The subtropical climate enables a particularly rich agriculture, with several harvests per year (up to three or even four crops a year). Rice is again the main crop, with fruits, sugar cane, tea, coffee, as secondary crops.

In spite of such an impressively diverse agricultural production, only about one-sixth of the total land area (almost one billion hectares) can in fact be cultivated, of which approximately 15.8 million hectares permanently support crops.

China's agricultural output is though the largest in the world. The country ranks first for the production of a number of commodities—among which rice, wheat, fresh vegetables, potatoes, watermelons, tomatoes, and pig meat—and its production sometimes far exceeds the one of the country which is ranked second (Fig. 1.1). In other major agricultural commodities, China usually ranks second (maize) or third (sugar cane, fruits).

Top Ten commodities Production quantity 2012
[1000 t]

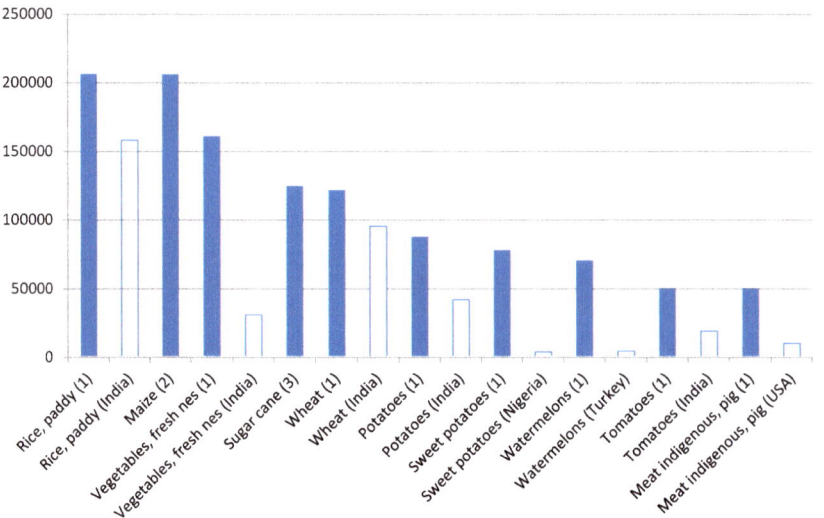

Fig. 1.1 China's top ten commodities production quantity and rank, and production quantities of countries ranked second when China is ranked first

The idea of this research was to select at least three case study areas, according to the following criteria: (i) economic conditions: from the least to the most developed provinces; (ii) levels of agricultural development: from the least to the most developed agricultural sectors; (iii) historical importance of governmental efforts dedicated to agriculture: from the most ancient efforts to the most recent ones; (iv) farming structures: from the least to the most modern farming structures (in terms of hectare per person, of level of integration in food chains, etc.). Accordingly, the following case studies were chosen:[3]

- Huangmo county, located in the arid part of Ningxia province, which ranks among the poorest provinces in terms of net income per capita in rural areas
- Lushan county, in Jiangxi province, better off than Huangmo in terms of net income per capita, but located in an hilly and inland area, where conditions remain difficult for agricultural moderniza-

tion although the area recently attracted attention from the government as the "abnormally underdeveloped" cradle of the Communist
Party
• Lanshui county, located in the inland part of Shandong province,[4]
suffering from delays in its economic development but with a strong
agricultural sector and investment capacities for further modernization

Each one of the earlier discussed case studies corresponds to one different zone in the "Three Rural Chinas" defined by Bernstein and Lü (2000:
241): "Industrializing rural China", "Middle-income agricultural China"
and "Low-income western China". Particular attention was paid not to
select "atypical" areas inside each great belt. For instance, Guangdong,
because of the importance of its political reforms, could be seen as an
atypical political area inside the Coastal belt. Tibet or Xinjiang, because of
the importance of religious and ethnic factors, could also be seen as atypical areas of the Western belt. Despite the fact that Ningxia is an autonomous region, the agricultural policies implemented in the area are not
much impacted by ethnical issues, as we will see later.

Fieldwork focused on a limited number of agricultural activities, namely
the production of fruits and vegetables. In Jiangxi, orange production was
investigated, and in Shandong, I focused on apple production.[5] Fruits and
vegetables are indeed important agricultural sectors, both in volume (712
million tons were produced in 2012[6]) and in the agricultural balance of
the country (the trade balance, for fruits and vegetables, exceeded US$10
billion in 2011, whereas the trade balance for cereals has been becoming
heavier and heavier in the past few years). However, these sectors suffer
from a lack of interest in the literature. Most of the scholars having done
research on agricultural production in China chose to investigate the grain
sector, which is seen as key for the food security of the country and, as
such (and as a sector which used to be heavily controlled by the state), was
specifically targeted by major policy reforms in the 1990s (Brown 1995;
Lyons 1998; Crook 1998; Aubert 1998; Zhou 1998) and in the 2000s
(Chen and Findlay 2004; Rozelle et al. 2000). In addition, fruits and vegetables constitute a highly interesting case study for research. The production systems and markets of fruits and vegetables were among the firsts to
be liberalized in the 1980s. As a consequence, as the overall trend of the
agricultural sector in China is marketization—the grain sector, which was
subject to the most stringent state control, was, in turn, liberalized at the
beginning of the 2000s—the current evolution happening in the produc-

tion systems and food chains of fruits and vegetables is likely to be representative of the future trends of evolution of other agricultural sectors. In addition, as the fruits and vegetables sector was liberalized very early compared to other agricultural sectors, it is most likely to include a wider diversity of stakeholders interacting with each other, from state officials to public *and private* enterprises as well as farmers of all sizes. In addition, contrary to the grain sector which is land-intensive, the production of fruits and vegetables is labor-intensive and there is a strong seasonality in production tasks with peak periods during treatment and harvest. As a consequence, the sector is likely to include a large number of diverse people taking part in production tasks at different periods of time, under different contracting models. For all these reasons, the sector offers an abundant and complex research material, particularly valuable to this research, which focuses on the sociological analysis of how rural stakeholders take part in and are affected by the new agricultural modernization.

A last case study was added in order to enrich the conclusions of this research on the pathway followed by China's agricultural modernization. A thorough exploration of "green" or "CSA" (Community-Supported Agriculture) horticultural farms in Beijing administrative area was conducted. These farms are indeed part of a relatively new form of agricultural enterprises and seem to belong to another agricultural modernization movement that the one that was observed in the earlier discussed case studies. In total, four case studies were thoroughly explored (Table 1.3 and Fig. 1.2).

These four case studies were complemented by a number of fieldworks conducted in other places, villages or investment zones, where I made observation and interviews. The following areas were explored: (i) one agricultural investment zone near Changzhou (Jiangsu), where I was invited by local officials; (ii) one village near Changsha (Hunan), where I lived with a local family of farmers (mainly growing rice and vegetables); (iii) three villages near Fengdu (Chongqing), where I investigated the activities conducted by an nongovernmental organization (NGO) working in the area on the improvement of maize productivity and on the development of small livestock farming; (iv) one village near Chaohu (Anhui), where I lived in a family of farmers (mainly growing rice and vegetables); (v) one dairy farm in Anhui province (Fig. 1.3). Although these areas were not thoroughly enough investigated to constitute "case studies", they contributed a lot to this research by providing additional material that was useful to check the conclusions drawn from the four

Table 1.3   Case studies

	Huangmo county (NINGXIA)	Lushan county (JIANGXI)	Lanshui county (SHANDONG)	Beijing CSA farms
Economic conditions	−−	−	+	++
Agricultural activity	Impeded by tough environmental and economic conditions	Traditional agricultural activity impeded by environmental and economic conditions	Traditional agricultural activity, strongly encouraged by the government	Depends
Governmental efforts toward agriculture	Somewhat weak	Somewhat strong although quite recent (2004)	Strong and ancient	Depends
Farming structures	Traditional	Traditional/modern	Modern	Innovative

main fieldworks. In addition, they provided elements to understand what was happening in other agricultural sectors, such as livestock farming or rice growing. Drawing on these preliminary elements as well as on secondary sources, it was possible to have a clear understanding of how other agricultural sectors were evolving in China under the modernization process, a comparative approach that contributed a lot to this research.

Although the idea was to "wipe the slate clean" and build categories according to interests, resources, and strategies of actors discovered along fieldwork, in line with Knoefpel et al.'s (2001: 47) definition of "empirical stakeholders", four preliminary categories of stakeholders were used to design interview outlines: government officials, enterprises, farmers, and NGOs. Semi-structured interviews were conducted with stakeholders from each of the four groups, with questions designed to elicit information on local patterns of relationships and to gather material on their resources, interests and strategies.

Although the fieldwork of this research relies heavily on qualitative semi-structured interviews, it is important to note that interviews were far from being the only source of information. Access to fieldwork in Chinese rural areas is difficult, but an important advantage compensates the trouble that one can have to get there: the fact that rural areas provide an incredible amount of directly observable information. Going to the fields during periods of peak activity enabled me to observe the number of farm-

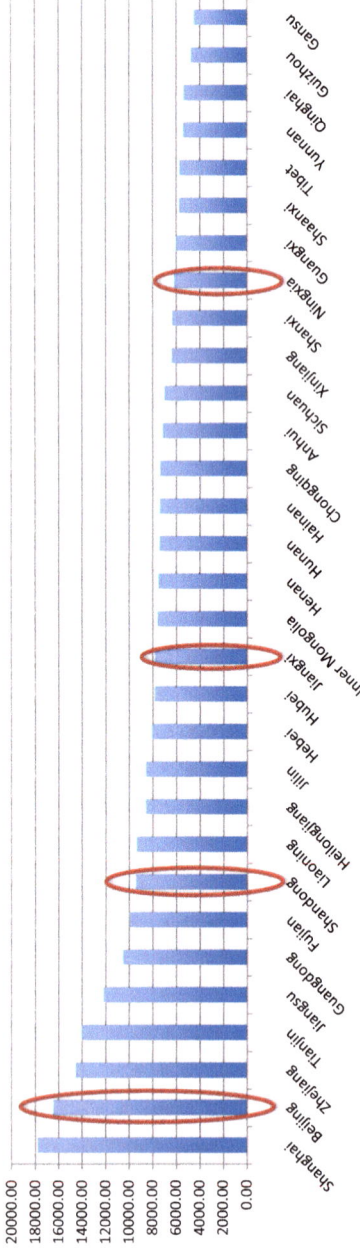

Fig. 1.2 Per capita net income per province (RMB 2012) (Source: Data from the National Bureau of Statistics of China)

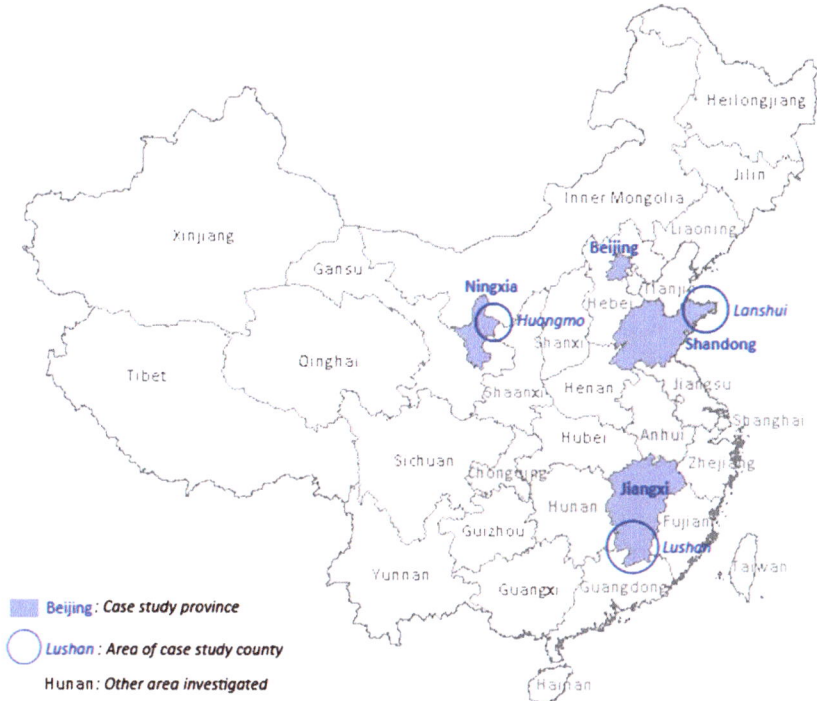

Fig. 1.3   Areas investigated for this research

ers working there and to exchange with them on their working conditions and daily lives, to check the presence of management staff in the fields, to observe their methods and farming techniques, to assess the quality of the products and so on. In addition, fieldwork also included visits of factories, which also offered a lot of directly observable information (e.g., the basic sociological profile of workers and management staff, working conditions, industrial processes, and traceability systems). Although most of the time, my main interlocutors were managers, I was able to cross-check their answers with information given by employees and workers.

The second fieldwork that this research relied onto focused on the defi-nition of the frames of reference of agricultural modernization at the cen-tral level of the government. In order to address this question, this research adopted the cognitive approach of public policy defined by Pierre Muller, for whom policies create "frames of reference" which play an important role in the shaping of collective action. According to the author, each policy has its own objectives and modes of implementation, which vary according to the *approach* adopted for the problem to solve. This approach, defined in political arenas, constitutes the frame of reference that puts order in a complex system of action. As Muller (2000: 1989) states it: "The purpose of public policies is no longer just to solve problems but to construct different frameworks for the interpretation of the world."[7] Muller's frames of reference are close to the "framework of ideas and standards" defined by Hall (1993), which designates "not only the goals of policy and the kind of instruments that can be used to attain them, but also the very nature of the problems they are meant to be addressing". However, Hall's paradigms, partly defined by exogenous factors (e.g., experimentation, social learning, and scientific circles), influence policymakers, whereas this research wishes to emphasize the normative capacity of public policy and its ability to build frames of reference and to shape collective action. To this end, the objective of this part of research was to acknowledge the existence of one or several framework(s), built by central level authorities to define a coherent frame for action for agricultural modernization in China. Fieldwork in local areas, on its side, provided elements to understand whether this (these) frame(s) of reference of agricultural modernization was/were effectively influencing the implementation of local policies in the countryside and whether local policies could, in turn, influence the frame of reference defined by central authorities.

The cognitive approach of public policy and the research on frames of reference are part of a relatively recent field of political science. Examples drawn from China were almost nonexistent at the time when this research was conducted. However, frames of reference are widespread for agricultural policies worldwide. The international debates arguing about which pathway "agricultural modernization" should follow—which have been particularly vivid over the past few years—are a clear sign that agricultural policies, today, are not just about implementing technical solutions to answer national demands but also define frames of reference that shape collective action for agricultural "modernization" (or "agricultural transition" in developed countries).

The aim of this step of research was to analyze the frames of reference existing in China for agricultural modernization. What are the key goals emphasized by agricultural policies? What are the key elements defining their discourse? Which tools, instruments, and stakeholders do agricultural policies promote? In order to provide answers to these questions, this research relied on the analysis of public policies and on qualitative interviews. In addition to the examination of first-hand sources (e.g., official documents and articles from Chinese media), interviews with key stakeholders were conducted to get a thorough understanding of the context and stakes at hand. Through this analysis, elements were gathered on stakes at hand, general and specific institutional rules, resources, stakeholders, political programs, action and implementation plans, and instruments. In a second stage of research, more in-depth interviews were conducted with targeted officials and researchers of the central level. Interviews aimed at assessing the validity of what was found in official documents (as the validity of data, and, in particular, the validity of national statistics, has long been the matter of intense debates in China (Holz 2002; Zhou and Ma 2005), particularly in rural China (Cai 2000)) and also crucial to gather information that did not exist in any document.

Interviews targeted high-level central officials—mostly from the Ministry of Agriculture (MOA)—and researchers and academicians close to central government authorities, working in natural sciences and social and political science in rural and agricultural institutes at Beijing and Shanghai's main universities and research centers.

For the selected group of respondents, qualitative semi-directive interviews were conducted, in English and Chinese, as well as participatory observation in exchange workshops. The interview outline was built in a way that could allow researchers and officials to express the "official" point of view (the one found in official documents) as well as their own point of view on current policies and on (alternative) solutions that (according to them) should be implemented to modernize the agricultural sector. The core of the interview guideline was made of questions linked to the role of stakeholders in the process.

1.4   Structure of the Book

Although the importance of the role played by social actors in the various phases of the policy cycle was evidenced and described by a vast body of literature in political science, social stakeholders still seem to be neglected by the policies aimed at triggering modernization of agricultural sectors—something that is not particular to China. Stakeholders are too often considered by policymakers as "rational economic actors", mostly driven by a willingness to increase their own profits. As this book wishes to emphasize, the wide array of stakeholders taking part in agricultural modernization are far from being rational actors instantly reacting to the implementation of agricultural modernization policies, and even less stirred by rationality and economic profit. Other dimensions are worth considering, such as path dependencies, institutional and cultural factors, and the established patterns of power and relationships in local areas. These dimensions greatly contribute to shape the frames of the transition pathway of the Chinese agricultural sector. Agricultural transition pathways are influenced both by the frame of reference promoted by the government through agricultural policies *and* by the action of stakeholders taking part in agricultural production, two objects this research focuses on.

The frames of reference promoted by agricultural modernization policies are always implemented on established social patterns. These local patterns of power and relationships and the capacity local stakeholders have to react to policy implementation greatly influence not only the efficiency of the implementation process itself but also the frames of the policy. As a consequence, the sociological analysis of the interests, resources, and power of the stakeholders taking part in agricultural modernization in a number of local places sheds light on how social actors, in turn, frame public objectives and political action. Having in hand the interests of stakeholders, a precise picture of local patterns of relationships in the concrete system of action of agricultural production and how they influence agricultural transition pathways, this book draws conclusions on the modernization pathway China is engaging on, bringing additional elements to the understanding of international debates on agricultural modernization trajectories.

In what ways do state actors restore their presence in agricultural production activities and what are the consequences for the agricultural modernization pathway China is engaging on? In order to address this research question, this books proceeds in several steps.

Chapter 2 serves as a historical introduction giving elements about the evolution of the interest of the Chinese state toward agriculture since the birth of the Chinese Communist Party (CCP). It identifies three main periods. First, it shows how the three aspects of rural life (peasants, agriculture, and the countryside) were key in the rise of the Communist Party during the Maoist era. In spite of the role they played, these areas, during a second period, were progressively relegated to the bottom of governmental priorities during the last two decades of the twentieth century, with urban and industrial development monopolizing the attention and consuming the largest share of government expenditure. The third period, which goes from 2004 to present, has witnessed a strong renewal of the state's interest in rural issues. The chapter provides an idea of the reasons why agricultural and rural development was put back in the agenda of the government.

Chapter 3 demonstrates how state agencies limit their ability to directly address the issues of inflation and food safety and preferentially rely on food-processing enterprises based in rural areas to modernize agricultural production. The chapter explains why these enterprises are the sole stake-holders really capable of addressing the issues of agricultural moderniza-tion, compared to NGOs or farmers. This chapter also investigates the recent (while limited) enlargement of this industrial and private-led agri-cultural sector to other private actors, essentially from downstream of the food chain (urban retailers) and, to a certain extent, from upstream of the food chain (agrochemical companies).

Chapter 4 relies on policy analysis and on interviews conducted both at the central level and in Shandong and Jiangxi to show how, in spite of the rising importance of private enterprises in the field of food production, state actors managed to keep control over this emerging agrarian entrepre-neurship. In particular, the analysis provides details on the formal and informal resources available to local government officials of county and township levels to increase their power over local entrepreneurs. The chapter also demonstrates that although state actors act as individuals steered by their own interests and preferences, a common framework of agricultural modernization, shaped by common goals and common tools, exists, is transmitted from the central level to local levels through various formal and informal channels, holds the state together, and enables offi-cials to act in a coordinated manner in spite of the fragmentation of the Chinese state.

Chapter 5 focuses on the often forgotten players of agricultural transi-tion: farmers. It describes the interests and strategies of small farmers, who often have no choice but to endure or escape their socio-economic situa-tion. The chapter depicts how the development of grassroots organiza-tions such as farmers' cooperatives or CSA was until now unable to empower small farmers and make them play a role in agricultural modernization.

Chapter 6 builds on the conclusions drawn in the previous chapters to characterize the pathway on which Chinese agriculture is embarking. In particular, the chapter points at particular institutional and social patterns, which are framed by policies and local players and prevent agricultural production to head toward more social and environmental sustainability.

NOTES

1. According to the National Bureau of Statistics, food expenditures still accounted for about 35 percent of urban and rural budgets in 2012, and could reach 43 percent for poor rural households (calculations done with data from the National Bureau of Statistics).
2. According to the National Bureau of Statistics of China, the average size of cultivated land per farmer is less than one hectare.
3. The names of the counties were replaced by pseudonyms to protect the identity of interviewees.
4. All counties were given pseudonyms, in order to protect our sources. Given that interviewees (individuals and companies) were sometimes selected among a small set of people and could be identified by their characteristics, it was indeed not sufficient to remove the names of these latest.
5. As we will see, it was not as easy as in Jiangxi and Shandong to find fruits and vegetables production areas in Ningxia that could have been interesting for this research. Therefore, in Ningxia, we had to focus on other types of products (but it did not change the content of our conclusions).
6. 140 million tons of fruits and 577 million tons of vegetables. As a comparison, 543 million tons of cereals were produced this year. Source: FAO database.
7. Original Language: *"L'objet des politiques publiques n'est plus seulement de 'résoudre des problèmes' mais de construire des cadres d'interprétation du monde."*

REFERENCES

Ash, R. F. (2006). Squeezing the peasants: Grain extraction, food consumption and rural living standards in Mao's China. *The China Quarterly, 188,* 959–998. https://doi.org/10.1017/S0305741006000518.

Aubert, C. (1998). The grain trade reforms in China: An unfinished story of state v. peasant interest. *China Information, 12*(3), 72–85. https://doi.org/10.1177/0920203X9701200304.

Barbier, M., & Elzen, B. (Eds.). (2012). *System innovations, knowledge regimes, and design practices towards transitions for sustainable development.* Versailles-Grignon: INRA-Département Sciences pour l'Action et le Développement (SAD). http://prodinra.inra.fr/record/174750

Bernstein, T. P., & Lü, X. (2000). Taxation without representation: Peasants, the central and the local states in reform China. *The China Quarterly, 163*, 742–763.

Bezes, P., & Le Lidec, P. (2010). Ce que les réformes font aux institutions. In J. Lagroye & M. Offerle (Eds.), *Sociologie de l'institution*. Paris: Belin.

Blecher, M., & Shue, V. (1996). *Tethered deer: Government and economy in a Chinese county*. Stanford: Stanford University Press.

Brown, L. R. (1995). *Who will feed China?* New York: W.W. Norton.

Cai, Y. (2000). Between state and peasant: Local cadres and statistical reporting in rural China. *The China Quarterly, 163*, 783–805.

Chen, C., & Findlay, C. (2004). *China's domestic grain marketing reform and integration*. Canberra: Asia Pacific Press.

Crook, F. W. (1998). China's "governor's grain bag policy": Concerns about food security. *China Information, 12*(3), 87–103.

Crozier, M., & Friedberg, E. (1977). *L'Acteur et le Système*. Paris: Editions du Seuil.

David, P. A. (1985). Clio and the economics of QWERTY. *The American Economic Review, 75*(2), 332–337.

Dorin, B., Hourcade, J. C., & Benoit-Cattin, M. (2013) A world without farmers? The Lewis path revisited. CIRED Working Papers, 47.

Haas, P. (1992). Introduction: Epistemic communities and international policy coordination. *International Organization, 46*(1), 1–35.

Hall, P. (1993). Policy paradigms social learning and the state. *Comparative Politics, 25*(3), 275–296.

Hall, P. (1996). *Governing the economy: The politics of state intervention in Britain and France*. New York: Oxford University Press.

Hjern, B., & Hull, C. (1982). Implementation research as empirical constitution-alism. *European Journal of Political Research, 10*(2), 105–115. https://doi.org/10.1111/j.1475-6765.1982.tb00011.x.

Holt-Giménez, E., & Shattuck, A. (2011). Food crises, food regimes and food movements: Rumblings of reform or tides of transformation? *The Journal of Peasant Studies, 38*(1), 109–144.

Holz, C. A. (2002). Institutional constraints on the quality of statistics in China. *China Information, 16*(1), 25–67. https://doi.org/10.1177/0920203X0201600102.

Kennedy, J. J. (2007). From the tax-for-fee reform to the abolition of agricultural taxes: The impact on township governments in north-west China. *The China Quarterly, 189*, 43–59.

King, F. (1949). *Farmers of forty centuries; or, permanent agriculture in China, Korea and Japan*. London: J. Cape.

Knoepfel, P., Larrue, C., & Varone, F. (2001). *Analyse et pilotage des politiques publiques*. Genève: Helbing und Lichtenhahn.

Lam, T. C. (2010). The county system and county governance. In J. H. Chung & T. C. Lam (Eds.), *China's local administration: Traditions and changes in the sub-national hierarchy*. London/New York: Routledge.

Lin, J. Y., Cai, F., & Li, Z. (2000). *The China miracle: Development strategy and economic reform*. Paris: Economica.

Lyons, T. P. (1998). Feeding Fujian: Grain production and trade, 1986–1996. *The China Quarterly, 155*, 512–545.

Mayntz, R. (1980). *Die Implementation politischer Programme In Implementaion politischer Programme Empirische Forschungsberichte I Empirische Forschungsberichte*. Königstein: Verlagsgruppe Athenäum, Hain, Scriptor, Hanstein.

Migdal, J. S. (2001). *State in society: Studying how states and societies transform and constitute one another*. Cambridge: Cambridge University Press.

Mollard, A. (2002). L'agriculture entre régulation globale et sectorielle. In R. Boyer & Y. Saillard (Eds.), *Théorie de la régulation, l'état des savoirs* (pp. 332–340). Paris: La Découverte.

Muller, P. (1984). *Le technocrate et le paysan: essai sur la politique française de modernisation de l'agriculture: de 1945 à nos jours*. Paris: Ed. Ouvrières.

Muller, P. (2000). L'analyse cognitive des politiques publiques: vers une sociologie politique de l'action publique. *Revue française de science politique, 50*(2), 189–208.

Nee, V., & Opper, S. (2012). *Capitalism from below: Markets and institutional change in China*. Cambridge, MA/London: Harvard University Press.

Oi, J. C. (1999). Two decades of rural reform in China: An overview and assessment. *The China Quarterly, 159*, 616–628.

Remick, R. J. (2004). *Building local states: China during the republican and post-Mao eras*. Cambridge, MA/London: Harvard University Asia Center.

Rozelle, S., Park, A., Huang, J., & Jin, H. (2000). Bureaucrat to entrepreneur : The changing role of the state in China's grain economy. *Economic Development and Cultural Change, 48*(2), 227–252.

Sabatier, P. A., & Jenkins-Smith, H. (Eds.). (1993). *Policy change and learning: An advocacy coalition approach*. Boulder: Westview Press.

Sabatier, P. A., & Mazmanian, D. A. (1979). The conditions of effective implementation: A guide to accomplishing policy objectives. *Policy Analysis, 5*(4), 481–504.

Steinmo, S., Thelen, K., & Longstreth, F. (1992). *Structuring politics: Historical institutionalism in comparative analysis*. Cambridge: Cambridge University Press.

Stone Sweet, A., Fligstein, N., & Sandholtz, W. (Eds.). (2001). *The institutionalization of Europe*. Oxford: Oxford University Press.

Swinnen, J. F. M., & Rozelle, S. (2006). *From Marx and Mao to the market : The economics and politics of agricultural transition*. Oxford/New York: Oxford University Press.

Tao, R., Liu, M., Su, F., & Lu, X. (2011). Grain procurement, tax instrument and peasant burdens during China's rural transition. *Journal of Contemporary China, 20*(71), 659–677.

Unger, J., & Chan, A. (1999). Inheritors of the boom: Private Enterprise and the role of local government in a rural South China township. *The China Journal, 42*, 45–74.

United Nations. (2011). *The global social crisis. Report on the world social situation 2011.* New York: United Nations.

Vanloqueren, G., & Baret, P. V. (2008). How agricultural research systems shape a technological regime that develops genetic engineering but locks out agro-ecological innovations. *Research Policy, 38*(6), 971–983.

World Bank. (2011). *Food price watch.* Washington, DC: World Bank.

Zhou, Z. (1998). Grain marketing systems in China and India: A comparative perspective. *Modern Asian Studies, 32*(2), 459–512.

Zhou, Y., & Ma, L. J. C. (2005). China's urban population statistics: A critical evaluation. *Eurasian Geography and Economics, 46*(4), 272–289.

CHAPTER 2

Agriculture: An Old Issue Back on the Public Agenda

2.1 THE ROLE OF THE THREE "RURALITIES" IN PARTY-BUILDING: REVOLUTIONARY PEASANTS, AGRICULTURAL REFORMS, AND EDUCATIONAL COUNTRYSIDE

Every nation has its founding myth. For Communist China, it is The Long March. (Sun Shunyun, *The Long March*)

China's agricultural and rural policies of the twenty-first century are enclosed in a specific framework, which was defined by the central government at the beginning of the 2000s under the name *san nong* (三农). In Chinese, *nong* (农) refers to agriculture, but also to "rurality" in the broader sense of the term, as *nongmin* (农民) means "peasants", *nongye* (农业) "agriculture", and *nongcun* (农村) "the countryside". A word-for-word translation of *san nong* could therefore be "the three ruralities". However, as *san nong* generally refers to the policy framework set up by the central government to address rural issues, it is generally translated as "the three rural issues". Despite the fact that this framework was developed in the 2000s, these three aspects of the rural life were determinant for the Communist Party way before the twenty-first century.

At the time of its official establishment, in late July 1921, the Communist Party counted no more than 50 members. After having experienced a

© The Author(s) 2018
M.-H. Schwoob, *Food Security and the Modernisation Pathway in China*, Critical Studies of the Asia-Pacific,
https://doi.org/10.1007/978-3-319-65702-8_2

steady then rapid growth in the years 1920s—its Fifth Congress, in April 1927, recorded nearly 60,000 participants (Yang 1990: 255)—the number of members suddenly dropped, just after the Nationalist Party launched its first campaign against the communists in 1927. The CCP was forced to retreat to the countryside, where communists found a fertile ground to expand their movement: exploited peasants. The pre-communist Chinese countryside was indeed under the domination of big landowners. Eighty-five percent of farmers were poor or middle peasants, owning only 37 percent of the national arable land. Only one-third of farmers had ownership rights over the soil they cultivated (Bouvier 1958: 95). Usury and high rents asked to peasants were impoverishing the countryside, already weakened by overpopulation and land fragmentation from generation to generation. According to Bouvier (1958: 95), the rent of bare land (without buildings, tools, or livestock) was reaching half—sometimes three-quarters—of the yield's value. As a consequence, peasants were often forced to resort to borrowing, with high interest rates. Conditions in which peasants were maintained were real seedbeds for anger and revolution. In addition, China already had a long history of rural uprisings, in which peasants were playing a leading role. In fact, peasants were often depicted as the central figures "in the rhythmic pattern of [the country's] millennial history" (Wilson 1971: 3). For Wilson, the fact that the leaders of the Communist Party were highly influenced by the heroes of peasant wars of the past was determinant in their strategy to look for the support of poor people in rural areas.

Finally, the Communists also jumped on the opportunity to fill a political vacuum. Yang (1990: 21) provides a particularly enlightening explanation of this "rural political vacuum" and on the strategic move of the CCP to make the best use of it: "For much of Chinese history, the rural society remained a domain independent of the state government and one offering various possibilities for peasant rebels, secret associates, local despots, bandits and warlords to challenge the government's authority […], possibilities [that] were turned—through the sophisticated agitations of the Communists—into the dazzling reality of mass revolutionary movement".

The Red Army, in exile in communist rural bases, needed to recruit people: at first, to regain military autonomy; then, to compensate for ever-increasing losses caused by the successive suppression campaigns launched by the Kuomintang (KMT). A lot of testimony exists on how cadres of the

CCP army were assigned recruitment targets and sent to rural areas sur-rounding the communist bases (Sun 2006). The essence of their discourse was fully in line with the objects of discontent of poor peasants. Their support was won on the basis of promises to end human exploitation per-petuated by landowners. During recruitment campaigns in "communist areas", promises were often translated into action. Sun (2006) provides examples of practices perpetuated in the red base of Jiangxi. He depicts how former rich peasants were granted the worst pieces of land, located on the side of hills or in marshy areas, whereas landowners did not have the right to own land anymore and were forced to be hired by others to sur-vive. Land and other goods previously belonging to landowners and rich peasants were redistributed to people supporting the Communists and, in particular, to the family members of new recruits.

Land reform really turned into a rallying cry for the Communist Party in the 1930s. In fact, as Yang (1990: 23) states it, "the entire decade from 1927 to 1937 was termed by the Communists the period of the Land Revolution, or more bluntly, the Land War". As Kerkvliet et al. (1998: 4) phrase it: "The war of liberation in China [was], notably, rural-based revo-lution". Kerkvielt et al. (1984) outline the difference with Russia, "the fount of Communist revolutions", where the Bolshevik Revolution "resembled more an urban coup than a protracted revolutionary strug-gle". Quite on the opposite of the CCP in China, the new Bolshevik gov-ernment was nurturing a suspicion of the rural areas and of the farming population, and "imposed collectivization almost as a war against the countryside". For Kerkvielt et al. (1984), this "suspicion of the peasant was entirely lacking in China [...], where, if anything, the villages were perceived as bastions of support for the revolution".

After five years of war between the Nationalist Party and the Communist Party, the Fourth Encirclement Campaign gave the first concrete results of Chiang Kai-shek's operations, wiping out two of the three major commu-nist bases at the end of 1932. The Fifth Encirclement Campaign forced the most important red base at that time, in Jiangxi, to engage, in turn, in a military retreat in October 1934. Historians generally take the end of the Fifth Encirclement Campaign as the beginning of the Long March. The "Long Marches", the military retreats of the nascent People's Liberation Army to evade the pursuit of the KMT, lasted until the spring of 1937 and involved tens of thousands of people. The retreat rapidly turned into a founding myth, in which the countryside played a tremendous part. The building of the myth of the Long March started even though the fleeing communists—among whom a lot were former peasants—were still fighting against elements and enemy troops running after them. Mao, leading the communist troops, started giving public speeches emphasizing the obstacles that the participants of the Long March had to victoriously overcome. The Chairman turned songs into hymns to the glory of the Red Army. He ordered the political department to gather stories of soldiers, among which 100 were selected and published in a book in 1938. Mao, in the end, managed to transform what had in fact been a military retreat in a glorious epic tale and what would become the spirit of the Long March.

The legacy of the myth of the Long March is still substantial today. An important number of high-level officials of the fifth generation of leaders—the current government—are the descendants of communist officials of the first generation, who took part in the early communist guerillas and in the Long March: they are known as "the princelings". Xi Jinping, for instance, is the son of Xi Zhongshun, who played an important role in the later stage of the Long March. In addition, the symbols inherited from the Long March are still used today by the Chinese officials. As Yang (1990: 1–2) phrases it: "The importance of the Long March can hardly be overemphasized, either historically or politically. Older Communist leaders have frequently referred to it as a turning point in CCP history; even now, fifty years later, survivors of the long march are still in control of China […]; and new Chinese leaders are calling their drive for economic modernization the 'New Long March'. The Long March has become a symbol of CCP history, just as the Great Wall is a hallmark of ancient Chinese civilization".

In spite of the important number of reasons that were given by historians to explain why the CCP "chose" to rely on peasants to start the communist revolution, the peasant base of the revolution was probably not entirely deliberate. According to Harrison (1972: 161–165), the first leaders of the CCP in fact wished to rely on the urban proletariat to lead China onto a "correct revolutionary road" and used to consider peasantry as elements of petty-bourgeois origin. In that sense, the original intent of the leaders of the CCP was close to the one of the leaders of the Bolshevik revolution in Russia. However, the "white terror" perpetuated by the KMT from 1927 on rapidly damped down the enthusiasm of factory workers for communism. The willingness of the CCP to rely solely on the urban proletariat to run the Chinese revolution was simply unrealistic (Harrison 1972) and would be equivalent to ignore four-fifth of the population (Bouvier 1958: 106). At the end of the year 1928, communist leaders started realizing how few ingenious the choice of neglecting rural areas was and progressively turned their interest toward the peasant movement.

For Harrison, the shift in the social base of the Party from proletariat to peasantry, "although, throughout its first fifty years, most leaders were in fact 'intellectuals'", was also partly due to the fact that the members of the CCP were pushed back by the KMT in the confined military controlled soviet areas, mostly settled in remote areas, deprived of industrial bases. The main consequence was that "while proletarians held certain leadership positions after 1927, the proportion of Party members who were of worker background fell from more than half in early 1927 to no more than 8 per cent in 1930, of whom less than 2 per cent were factory workers" (Harrison 1972: 148).

Peasants rapidly started forming the major part of the CCP's army. In April 1934, just before the March, they would have constituted 68 percent of its ranks. Proletarian workers, on their side, would have accounted for only 30 percent of the communist military forces at that time (Wilson 1971: 70). The Chinese revolution turned into a peasants' revolution, conducted by a leader coming from a family of farmers from Hunan.

In 1949, the Communist Party came to power and put an end to the domination of big landowners by redistributing land property rights to poor farmers. The scale of redistribution was colossal: 47 million hectares—46 percent of the cultivated area—were distributed to 70 million peasant households, who received a little over half a hectare per family (Bouvier 1958: 95). Right after this redistribution, a national-scale collectivization program for land and agricultural resources quickly grew in the mind of communist leaders. For Ngo (2009: 285), the collectivization "represented a critical stage in the Chinese Communist Party's state-building", because cooperatives were a way to link the state to villages. The transition to collectivized agriculture took place gradually. At first, "mutual-aid teams" were created, at the beginning of the year 1952. Membership was mainly on a voluntary basis, but as mutual-aid teams provided their members with significant advantages (e.g., financial and technical support from the government), they rapidly aroused the interest of rural households. In 1953, the first agricultural cooperatives per se were created. From "elementary cooperatives" gathering small groups of peasants on small areas and running under a "semi-socialist" system (the work of each farmer was rewarded according to his amount of effort and participation to agricultural tasks and according to the capital brought to the cooperative), the system evolved to "advanced cooperatives", usually gathering around 250 households. Under this latest system, land and tools were fully owned by the collectivity, and members were entirely paid according to the rules of a "work points" system. In spite of a certain resistance of peasants facing the collectivization of their goods (Ngo 2009; Li 2009), transition from elementary to advanced cooperatives was successfully carried out. At the end of the year 1956, almost 90 percent of rural households were members of advanced cooperatives (Li 2009: 39).

In August 1958, at a conference in Beidaihe—the summer residence of leading government dignitaries of the Communist Party—the Central Committee adopted the new designation of "People's Communes", and made them part of the "three great banners", along with the new program for building socialism and the Great Leap Forward. Communes were much larger than advanced cooperatives, as a single commune could count several thousand rural households. They were organized according to a hierarchy of administrative entities. Each commune was organized in brigades, which were in turn divided in production teams. At the end of the year 1958, the Chinese countryside was divided in 26,000 communes.

People's Communes radically changed the agricultural production model. Local officials were put in charge of production and could make decisions in terms of task allocation and working-time distribution. In the course of the progressive establishment of cooperatives, this new distribution of power gave rise to debates. Villagers were sometimes reluctant to give back the land they were granted with when the CCP had come to power. However, cooperatives were an essential tool to control agricultural activities and people in rural areas, and for this reason, debates were rapidly cut off. As Bowie and Fairbank (1962: 4) put it: "The cooperative farm system made it easier for the Party to control labor and to collect grain taxes. It was no doubt for this reason that Mao insistently opposed the indiscriminate dissolution of agricultural producers' cooperatives". The system of collectivist agriculture was running along with a system of work points, which were granted to farmers proportionally to the time spent working in fields. Work points allowed workers to have access to a proportional quantity of food in mess halls. The control of the basic needs of rural residents became a powerful domination mechanism for local leaders, especially when times of food shortage came.

In the 1950s, the Communist Party not only completely rethought and reorganized agricultural production into collective farms. The leaders of the CCP also gradually established a nationally planned system for the production and distribution of agricultural products. Agricultural production was essential to sustain urban growth. Urbanization rate had already jumped from 10.64 percent in 1949 to 15.39 percent in 1957 (Chen 2008: 8) and urban population had swollen more than 100 million people. Grain consumption had kept on rising accordingly. In order to answer the rise in urban food demand, productivity targets were assigned to local officials in the countryside. Objectives were decided at the central level and promulgated through Five-Year Plans. As an illustration, the first Five-Year Plan (1853–1957) set up the following national objectives (Bowie and Fairbank 1962: 54–55):

> The First Five-Year Plan sets suitable targets for increased agricultural outputs. [...] According to the plan, the projected output of staple farm products for 1957 and the expected percentages of increase over 1952 are as follows: Grain: 385,600 million catties,[1] an increase of 17.6 per cent. Cotton: 32,700,000 *tan*,[2] an increase of 25.4 per cent. Jute and ambary hemp: 7,300,000 *tan*, an increase of 19.7 per cent. Cured tobacco: 7,800,000 *tan*, an increase of 76.6 per cent. Sugar-cane: 26,300 million catties, an increase of 85.1 per cent. Sugar-beet: 4270 million catties, an increase of 346.4 per cent. Oil-bearing crops: over 118 million mou will be sown, an increase of 37.8 per cent over the acreage of 1952.

The State Planning Commission, established in 1952, played an important role in the implementation of the first Five-Year Plan. The Commission was relying on a network of ministries and local planning bureaus. Whereas the most important targets were set by the highest levels of the government, ministries were in charge of targets for the commodities considered as less important for the national economic growth, and local planning bureaus were in charge of the implementation of the plan.

Grain was ranged straightforward among the most important commodities and, as such, was rapidly imposed governmental control. In 1953, a state monopoly on grains was decreed[3]: all surplus grain had to be sold to the state at fixed prices. State granaries mushroomed and quotas per head were established. At that time, grain still constituted the greatest share of the food ration. National planning of food distribution aimed at ending speculation and stabilizing the price of basic staple products.

Planned economy gradually became the rule for other commodities as well. Private markets closed down and state goods, produced by enterprises mandated by the state, started being sold instead at fixed prices. Cities were granted priority in the distribution of grain. In 1950, the government, faced to the risk of a decrease of the farming workforce consecutive to the rise in urban population, established a national household registration system (户口 *hukou*), which divided the population in two categories: agricultural and nonagricultural population. This system considerably limited rural-urban migration and created the roots of a strong urban-rural divide, which caused important inequalities that remain among the contemporary Chinese society even today. In order to meet the growing urban demand, agricultural products were plucked out of the countryside, on the basis of production targets which were sometimes established on inflated yield figures by local officials finding an opportunity to earn merit through such a process (Dikötter 2010: 37).

Ambitious food production targets were also established at the national level in order to honor export contracts with foreign countries. The refusal of Mao to cut on exports—against the opinion of other CCP leaders—is considered by Dikötter (2010) as one of the most important factors that led to the Great Famine of 1958–1961 (before the end of the year 1960, when Zhou Enlai and Chen Yun, close advisors of Mao, finally managed to convince the Chairman that grain had to be imported from foreign countries). To this day, the pressure of the Soviet Union to pay back debts is still considered by many as the main cause of the Great Famine, along with natural catastrophes.

Assigning production and export targets was supposed to enable China to import industrial products, and was a way to provide raw material and a suitable ground to industrial revolution, considered at that time as a major pillar of the economic "catching up" of China. In the 1950s, the whole economy was relying on agriculture. The agricultural sector provided 90 percent of the raw material for consumer goods industries, and industrial imports were paid thanks to exports, of which agricultural products represented 75 percent (Bowie and Fairbank 1962: 3). In the words of the CCP: "The great tide of agricultural co-operation that has swept China is bringing forth an immense, nation-wide growth of agricultural production, and this in turn is stimulating the development of the whole national economy" (Bowie and Fairbank 1962: 120). Agricultural development was considered as an essential first step toward economic power. Economic power gained from agricultural and industrial development was then supposed to lead China to be one of the world's leading political powers. The Chairman did not only wish to radically change the system through the carrying out of socialization in economic sectors. He also longed for a transformation of people's minds. In this process of ideological and sociological remodeling, the countryside played a major role. In order to lessen the status and influence of intellectuals and to ensure the spreading of the proletarian leadership, intellectuals and young people were sent down to rural areas in order to be "reeducated" by workers, peasants, and soldiers. The sending of young people and intellectuals to the countryside started as soon as the Communist Party came to power. As mentioned by Chen (1974: 95): "As early as 1945 Mao Tse-tung said that intellectuals 'should gladly go to the countryside, put on coarse clothes, and willingly take up any work, however trivial'. [...] At different times in the first decade of the regime, students as well as more mature intellectuals were urged to go to the rural areas to take part in agricultural production".

Sending urban dwellers to the countryside was not only serving ideological purposes. It also aimed at slowing the growth of the urban population—of which food demand kept on rising—and at increasing the number of people working in the agricultural sector. According to the China Development Research Foundation (2013: 13), a great number of people were "sent down" in the aftermath of the Great Leap Forward, as a way to curb the rise in food demand and as a way to ease food shortage:

> The failure of the Great Leap Forward, together with natural disasters, forced the country to carry out adjustments to the national economy. The super-fast increase in the urban population had clearly exceeded the capacity of grain supply at the time. Starting in 1961, a large-scale effort began to reduce the urban population in order to mitigate famine. Urban population were 'mobilized' and returned to rural areas. The urban population was reduced by roughly 20 million in the two years of 1961 and 1962. The urbanization rate declined from 19.8 per cent in 1960 to 14.6 per cent in 1964. Only in 1965, by which time the national economy had basically recovered, did it rebound to 16.8 per cent.

The movement of sending people down to the countryside reached its peak at the end of the 1960s, during the first years of the Cultural Revolution. At this time, the process of sending people down had gone back to its original ideological purposes. As Chen (1974: 95) puts it:

The [1968] campaign to send the intelligentsia to the countryside surpasses all previous efforts in proportion and scope. There is now a broader meaning in reeducation: not only do the intellectuals who are the products of bourgeois education need to be reeducated by the laboring class but the young people who attend schools dominated by intellectuals [...] must be purged of the ill effects of the wrong kind of education. [...] Estimates [of the total number of people sent to the countryside since the stepped-up campaign of 1968] vary from 25 to 30 million to 40 to 60 million.

In the countryside, workers and peasants were supposed to teach young people and intellectuals the "simple virtues" of peasant life, filled with hard work and unburdened of the luxuries characterizing urban lifestyles. During the times of the Cultural Revolution, even though the main purpose of the movement was ideological, the issue of providing human resources to supply the needs of agricultural production in terms of labor was still pending. Employment and other problems had also started to emerge in cities, and sending people down to the countryside was another way of thinning out urban population and of solving rising urban issues.

According to Mao's thinking, ideological remodeling was necessary and hard work was key in the process. The program of sending people down to the countryside did not solely aim at rectifying the mind of "deviant" elements or punishing people resisting revolutionary ideas. Propaganda teams and the Communist Youth League were actively trying to convince people to send their children for "rural service" for their own benefit. Tens of millions of people were sent down to rural areas, either temporarily or permanently. In just four years, between 1968 and 1972, around 42 million "educated youth", cadres and other urban dwellers were "sent down" to the countryside (China Development Research Foundation 2013: 13).

Workers, soldiers and peasants, were considered as allies of the Communist Party and viewed as key players in the process of socialization. In the trio worker–peasant–soldier, peasants, because of their demographic weight and because of their opposition to the "traditional bourgeois elite of intellectuals", were considered to be the most dynamic revolutionary force once the CCP had come to power. However, hard work and rural lifestyle were not enough to re-educate urban masses, and a role was also given to lower and lower-middle peasants. As Mao stated it: "It is necessary for educated youth to go to the countryside to be re-educated by lower and lower-middle peasants" (Chen 1974: 273). Farmers were entrusted to teach values to young people and intellectuals coming to the countryside. In some areas, they were also given control of rural schools.

Giving a role to lower and middle-lower peasants was also a way of controlling local cadres. Agricultural collectivization indeed came along with the creation of millions of grassroots cadres, who were soon given considerable power over peasants. As Li (2009: 5) suggests:

> to discipline the cadres, the state could only rely on the initiatives of ordinary people through two means: the imposition of various institutions that allowed the 'masses' (*qunzhong*) or ordinary people to supervise the cadres from the bottom up, and the making of a new discourse that empowered the masses by assuming the political correctness of the 'poor and lower-middle peasants' (*pingxiazhongnong*) and their supremacy on the corruptible cadres.

Peasants, first as a revolutionary force and then as the "guardians" of the values promoted by the Maoist ideology, played a fundamental role in the building of the CCP, from its earliest times to the end of the Cultural Revolution. Rural areas, by offering a refuge against the KMT campaigns, a political vacuum to expand the power of the CCP, and scenery for the founding myth of the Long March, also form a major part of the collective psyche of the Communist Party. Finally, agricultural development through Communes, as the first step of the economic catching up of the great power to come, was also among the pillars of the building of the legitimacy of the CCP. However, in spite of the place peasants, rural areas and agriculture had in the Party-building discourse from the 1930s to the 1970s, in the late twentieth century, these "three aspects of rurality" had cruelly lost the interest of the government.

2.2   AGRICULTURE AND THE STATE IN THE LATE
TWENTIETH CENTURY

The decrease in the interest of the government for rural areas in the 1980s and 1990s is visible on several items: during this period, the share of central expenditures dedicated to rural areas shrank and the documents produced by the central government barely mentioned rural issues anymore—despite, as we are about to see, strong central administrations in charge of rural policy. At the local level, institutional capacities to implement reforms in the agricultural sector severely weakened after the implementation of institutional reforms and village elections, affecting even more the state capacity to stir agricultural development.

Overlapping responsibilities, which are regularly depicted by political scientists as a special feature of the Chinese government (Feng et al. 2006; Vermeer 1998), are usually considered as a legacy of the Maoist era. In 1970, China had indeed more than 100 ministries and commissions (Waldron et al. 2006: 282) competing for economic gain and political power. In the 1980s, after the arrival of Deng Xiaoping to power, the government initiated a fundamental transformation of its administrative system. The main goal of this transformation was to transfer to enterprises the productive functions of the economy, which were formerly achieved by governmental institutions. Reforms pushed the state to give up on mechanisms directly controlling the economy, which were supposed to be replaced by less direct macro-level control mechanisms such as subsidies or loans, allowing governmental institutions to keep on steering economic development. Government bodies in charge of machinery industry, metallurgical industry, light industry, textile industry, and so on were abolished, demoted, or merged, giving birth or elevating government bodies such as the ones in charge of development and reform, industry and commerce administration and so forth.

At the beginning of the 1980s, governmental institutions had first benefited from an increase in the number of state employees—particularly in the fields of the economy linked to development, such as infrastructures or education. However, reforms rapidly led to a serious downsizing of public institutions. From 1999 to 2002, in just three years, the number of state employees dropped from 83 million to 69 million people. Personnel reductions started addressing the overlap of responsibilities, which was particularly acute at the end of the Maoist era (Waldron et al. 2006: 284).

The Ministry of Agriculture was put through personnel reductions as well. Between 1990 and 2002, the number of employees in charge of agricultural issues was almost cut by half, dropping from 7.3 million to 4.1 million (Waldron et al. 2006: 280). A number of functions previously carried out by the ministry were transferred to other central state departments. However, in essence, administrative reforms left the power of the Ministry of Agriculture relatively unimpaired. In the middle of the 2000s, officials working on topics related to agriculture indeed outnumbered by far state employees working in other sectors. This comparison made by Waldron et al. (2006: 280–281) gives a clearer idea of the situation: "The number of state staff in agriculture is comparable to service sectors such as health, sports and social services (combined), and transport, storage, and post and telecommunications (combined) and only overshadowed by the education sector".

At the central level, the Ministry of Agriculture was maintained and kept on working directly under the State Council, proving that the agriculture was not considered as just one sector among others—like textile industry—and was regarded as a key sector deserving dedicated central institutions. Administrative reforms of the 1990s and 2000s, far from having weakened governmental institutions in charge of agricultural reforms, seem, on the opposite, to have granted greater power to the ministry, relatively to other central institutions.

The corollary of the minor effect of reforms on the agricultural administration is the resulting persistent issue of overlapping responsibilities. Despite the fact that the issue is clearly not unique to China (similar coordination problems affect a large number of countries in a wide variety of political fields, from Japanese foreign policy (Ahn 1997) to sanitary crisis in Great Britain (Greer 1999)), a body of evidence in the literature suggests that overlaps represent a strong feature of the Chinese administrative system, with the theory of fragmented authoritarianism (Lieberthal 1992), for instance, underlining a number of structural difficulties preventing the government from ensuring efficient coordination between the local administrative entities of the Chinese system. Interviews I conducted regularly mentioned overlaps as an important issue impeding the effective implementation of agricultural policies. Several governmental bodies indeed take part in the decision making of public policies related to the agricultural sector. The Ministry of Agriculture is officially in charge of designing middle and long-term strategies, politics and programs aimed at developing agriculture and rural areas. In addition, it also has to organize and supervise the implementation of these programs and policies. Finally, the ministry can draft legislation related to agriculture, agricultural inputs, and rural industry—on which the National People's Congress and its Standing Committee have the final decision. Performing such tasks can be difficult in an environment where resources essential to agricultural production are managed by other ministries—such as the Ministry of Water Resources or the Ministry of Land and Resources. In addition, the responsibilities of the Ministry of Agriculture are likely to overlap the ones of other institutional bodies. For instance, the Ministry of Agriculture has to "revitalize agriculture through science and education". Such a mission includes the management of scientific and technological research programs, which infringes upon the tasks of the Ministry of Sciences and Technology. It also includes the handling of agricultural education, which might overlap the responsibilities of the Ministry of Education. Many other examples could be given (with the Ministry of Commerce, the Ministry of Foreign Affairs, etc.).

Overlapping responsibilities also exist between the Ministry of Agriculture and other ministerial bodies, which are not granted the name of "ministry" but are nevertheless at ministerial level (e.g., the Administration for Quality Supervision, Inspection and Quarantine (AQSIQ), of which the responsibility in food safety affairs is regularly pointed out, such as when in 2008, the melamine milk scandal led to the resignation of the AQSIQ chief, Li Changjiang), or between the Ministry of Agriculture and other nonministerial but powerful bodies under the National Development and Reform Commission (NDRC) (e.g., the State Grain Administration, which is in charge of controlling national grain distribution, of drafting guidelines for grain industry, and of managing national grain reserves).

In addition, because of the always-stronger link between urbanization and agricultural and rural development, functions previously assumed by the Ministry of Agriculture increasingly need to be coordinated with the action of the Ministry of Housing and Rural and Urban Development. The growing stakes of environmental issues and their obvious connection to agricultural activities—agriculture consumes more than 60 percent of the water resources of the territory (China Water Risks 2014) and emits large quantities of greenhouse gases—also creates an urgent need, for the Ministry of Agriculture, to establish strong links with the Ministry of Environmental Protection.

In practice, barriers prevent effective communication and coordination between the different administrations of the central state. These barriers are not unique to agricultural issues and can be found in other political fields as well. For a number of issues, transversal commissions have been established in order to coordinate the activities of various governmental bodies on a specific subject. According to Yu (2008), for instance, the setting up of the National Coordination Committee on Climate Change significantly improved the Chinese answer on the issue, both nationally and in international forums. In sectors linked to agriculture, however, coordination usually remains weak.

To sum up the above, the administrative reforms of the 1980s let the power of central agricultural administrations relatively unimpaired—but at the same time did not solve the issue of overlaps in responsibilities for the design of agricultural policies. Although the Ministry of Agriculture was deprived from its capacity to plan agricultural production through the People's Communes at the beginning of the 1980s, the central structure was then relatively spared from the personnel reductions of the administrative reforms comparatively to other Ministries. However, the fact that the Ministry of Agriculture was still strong in the 1980s and 1990s did not help it in putting rural and agricultural development on the agenda of the central government overall. Starting from the middle of the 1980s, the priorities of the central government shifted to industrialization and urbanization. It is clearly observable in the Five-Year Plans of this period. The main principles of economic development recommended by the Seventh Five-Year Plan (1986–1990), for instance, put strong emphasis on industry and science and technology but do not mention agriculture. Among other things, the plan insists on the necessity to adjust the industrial structure to the changing needs of the population,[4] on the need to accelerate the building of the energy sector, the transport and communication sector, and the raw and semifinished material production sector,[5] and on the need to increase efforts in the development of science and technology. The Eighth Five-Year Plan (1991–1995) and the Ninth Five-Year Plan (1996–2000) are more explicit on the importance to develop agriculture. However, in the 1990s, no significant agricultural reform was conducted apart from the ones affecting the grain sector.[6]

In addition, most of the financial effort made by the government during the second half of the 1980s and during the 1990s was dedicated to the development of the industrial sector and urban areas. An unbalance of expenditures progressively appeared in the 1980s and started disfavoring rural development. During this period, however, the government in fact increased the amount of expenditures dedicated to rural development. For instance, investments for the building of new irrigation systems rose from 10 billion RMB in 1978 to 43 billion RMB at the end of the 1990s (Bruins and Bu 2006: 117). Education also benefited from an important increase in rural budgets: at the end of the 1990s, public funds aimed at improving education in the countryside reached 48 billion RMB, compared to 10 billion in 1978 (Yu and Zhao 2009: 11). Government expenditures on agricultural production and administration rose too, going from 10.1 billion RMB in 1985 up to 22.2 billion in 1990 and 43 billion in 1995.

However, despite the rise in absolute government expenditures allocated to rural areas, their share in the national budget decreased, in favor of investments allocated to urban areas and to the industrial sector. From

7.6 percent of the GDP in 1978, public agricultural investments fell to 3.6 percent in 1995 (Huang and Rozelle 2009: 19). Østergaard (1990: 9) gives another useful data that show the shrinking of public investment in agriculture in the 1980s: "Government expenditures on agriculture as a percentage of total expenditures decreased from 13.7 in 1979 to 8.1 percent in 1988. Over the same period, State capital construction funds invested in agriculture declined from 11.9 percent of total construction funds to just 2.9 per cent". This unbalance rapidly entrenched economic and infrastructures inequalities between rural and urban areas.

In addition to the shift in central government priorities, rural areas were also impacted by a number of reforms, which further confirmed the waning interest of the Chinese state in these issues, which became obvious at the local level as well. The weakening of local state authorities in charge of agricultural production first started with the abolition of People's Communes. The planning of production and distribution of agricultural products, set up in the early years of the CCP, had shown important weaknesses. The peasants' loss of control over their working time and cultivation patterns had led to a significant decrease in agricultural output (Li 2009: 49). Collective property of agricultural tools and machinery were giving little incentive to farmers to take good care of them. The work points system had encouraged peasants to focus on the amount of time spent in the fields rather than on work efficiency. Conflicts had arisen between farmers, facing the practical impossibility of escaping their situation giving the rigidity of the *hukou* system, and local officials, who had to cope with important pressures from above, being held accountable for the amount of grain sent to cities. Finally, the mistakes of agronomic programs implemented at the national level and replicated at the Chinese scale had disastrous consequences on yields. To name just a few: deep seeding depleted soils, close seeding choked out plants, and extermination campaigns of birds led to the development of worms population.

The inefficiencies of a planned distribution at the Chinese scale had rapidly threatened food security in many areas. State employees were unable to handle the buying, stocking, transport, and distribution of grain at the national scale, which led to a serious situation of both food shortage and food waste, even though a number of state granaries were filled with grain (Dikötter 2010).

In parallel of the reforms conducted in the agricultural sector, tremendous changes occurred in the industrial sector under the Maoist era. Colossal targets were set. Steel production was supposed to jump from

5.35 million tons in 1957 to 12 million tons in 1960, and to reach 100 million tons in 1962 and 700 million tons in 1975 (Dikötter 2010: 57–58). Small furnaces mushroomed in the countryside and numerous agricultural tools ended up feeding their fire.

The combination of all these elements resulted in the Great Famine of 1958–1961, which was responsible of tens of millions of deaths in just three years (estimates vary from 20 million (Aird 1982) to 45 million deaths (Dikötter 2010)). In the years following the Great Leap Forward, the decision to raise food imports eased the situation. However, agricultural production took time to recover and agricultural output started increasing again at a very slow pace only in the middle of the 1960s (Huang and Rozelle 1997: 339).

In 1979, soon after Deng Xiaoping's arrival to power, fundamental reforms were implemented. The new de facto leader of the People's Republic of China (PRC), in line with the "Four Modernizations" policy enlightened by the failures of past experiences, radically changed the agricultural production system. Reforms started with the abolition of agricultural collectivization. People's communes were progressively dismantled and land was reattributed to rural families, which were "given" small plots of less than half a hectare. In practice, rural families rent land, which is formally owned by village committees. The duration of the leasing contract, in the early years of the dengist reforms, was set at 15 years.[7]

In parallel to land redistribution, the "Household Responsibility System" (HRS) was established. Rural households regained the complete control of cultures and farming methods and from then on, agricultural profits entirely went back to farmers. The HRS rapidly proved efficient and an important number of areas quickly adopted the system. From only 5 percent in 1980, the proportion of communes running under the HRS jumped to 67 percent in 1982 and reached 98 percent at the end of 1983. The fact that the income generated by land and farm work would from then on entirely benefit farmers was a strong incentive for these latest to look for productivity gains, to turn to more cost-effective cultures and methods and to maintain land and tools in good conditions. Consequences on production were substantial: grain productivity surged from 2527 kg per hectare in 1978 to 3608 kg per hectare in 1984 (Bruins and Bu 2006).

Agricultural markets were also gradually liberalized. In 1985, the number of products of which markets were directly controlled by the state had been reduced by two thirds. During the second part of the 1980s, liberalization spread to a wider range of products such as pork, fish, chicken, tea, and fruits and vegetables. The rapid growth of the urban population—the urbanization rate goes from less than 18 percent in 1978 to more than 23 percent in 1985 and to almost 30 percent in 1997—along with economic development[8] stimulated the demand for more diversified food products. Market liberalization and the diversification of demand freed farmers from their former obligations to produce more grain. Farmers, who used to work to fulfill grain quotas required by local production teams, were from then on able to turn to other products. Consequences on agricultural diversification were tremendous. From 1978 to 1990, surfaces dedicated to commercial crops almost doubled (Ash 1992: 570). Farmers chose to turn to cash crops, but also gave up on grain farming to concentrate on livestock and aquaculture. The share of livestock farming and aquaculture in the agricultural value added went from 15.5 percent in 1978 to 25.8 percent in 1990. Between 1981 and 1985, pork, beef, and mutton production average annual growth rates were close to 10 percent. The development of aquaculture production was even more impressive: 9.4 percent annually (Ash 1992: 548) between 1981 and 1985, and 13.7 percent annually between 1985 and 1995 (Huang and Rozelle 1997).

According to a number of scholars, most of the rise in agricultural productivity at the beginning of the 1980s can be attributed to the instauration of the HRS. At that time, farmers indeed started paying more interest to cropping choices and farmwork, as profits made from productivity rises directly went in their pockets. Whereas the annual rate of increase of agricultural production was about 7.1 percent during the years following the establishment of the HRS, Huang and his team (Tongeren and Huang 2004: 35) acknowledge a slowdown in the growth of agricultural production once the effects of the institutional reforms had been harvested. As the researchers state it: "As by the mid 1980s the one-off efficiency gains from the shift to the household responsibility system (HRS) essentially had been reaped, the growth rate of the food and agricultural sectors decelerated".

The abolition of the Communes and the establishment of the HRS enabled farmers to turn to economically more attractive agricultural activities, which had considerable effects on their income. Between 1978 and 1985, the revenues of rural families, in average, grew by 15 percent annually, and the net revenue per household more than doubled, going from 134 RMB per year in 1978 to 398 RMB in 1985 (Li et al. 2006: 15).

Whereas collectivization had deprived farmers from their ability to make agricultural production choices, the dismantlement of cooperatives, the instauration of the HRS, and the liberalization of markets restored their responsibilities and control of agricultural production and considerably reduced their former dependency toward local officials. Collectivization had indeed not only been about state control over rural economic activities such as agricultural production. Agricultural collectivization had also had a tremendous effect on the pattern of relationships between rural dwellers. Millions of grassroots cadres were charged with the responsibility of managing rural affairs. They were granted with considerable power over peasants, as they were allocating work time, giving peasants work points according to the amount of time spent in the fields, and controlling communal canteens. As Li (2009: 5) sums it up, "state penetration of the village [had] reached an unprecedented level during the collective era". The establishment of the HRS put an end to these domination mechanisms and completely changed the pattern of relationships between peasants and local state officials, well beyond the sphere of daily agricultural production activities.

The road to the greater independence of farmers did not end with the dismantlement of the People's Communes and the establishment of the HRS. Political leaders also quickly started thinking about granting villages with the possibility of governing themselves. Self-government at the village-level was proposed by the central government as early as in the beginning of the 1980s. The n°111 article of the 1982 Constitution defines "village committees" (村民委员会 *cunmin weiyuanhui*) as "self-governing organizations of farmers at the lowest level" (Li 2009: 292). Local governments progressively endorsed the reform and village committees spread across rural China. In 1984, there were already around one million village committees throughout the whole country (Li 2009: 292). In 1987, the Organic Law of Village Committees was issued in order to establish direct elections in villages for village committee members. At first launched on a trial basis, the law was fully adopted by the National People's Congress in 1998. Villagers aged 18 years and above could from then on elect the members of village committees every 3 years.

The establishment of village committees and of direct elections for committee members was supposed to give self-government rights to villagers. However, during the period following the reform, the most important functions—the ones related to economic development or to the salary of officials—as well as the power to take final decisions remained in the hands of the Party secretary, who kept an important role in the management of local affairs. The law was revised in 1998 and granted village committees with new powers, such as the collection of fees, the raising of funds, and the management of land and other resources. Even if the Party secretary, at the village level, sometimes still plays a key role in the handling of the village's affairs,[9] village committees and direct elections considerably changed relationships between local cadres and farmers.

In addition, the change in the political leadership significantly lowered the importance of propaganda in rural areas. Mass meetings and group studying, which were common under the Maoist era, disappeared, both because means for exerting pressure over villagers were withdrawn from local cadres and because these latter, from now on evaluated on economic and social stability criteria and on the results of one-child policy they had to enforce, had no interest in keeping on convening ideological meetings anymore. This led to a real depoliticization of the countryside. According to Huaiyin Li, the retreat of the state enabled traditional ties to revive. Because peasants could not rely on production teams anymore whenever encountering problems related to agriculture, they started turning back to family members for mutual help in the fields or to borrow money (Li 2009: 305).

In rural areas, cadres progressively turned their attention toward the development of industrial activities. The number of *agricultural* township and village enterprises (TVEs) dropped from 495,000 in 1978 to 231,000 in 1991, whereas in the same period of time, the number of *industrial* TVEs increased from 794,000 to 7,426,000 (National Bureau of Statistics Database). At the end of the 1990s, the role of the state had considerably decreased in agricultural activities, which were mostly taken care of by farmers. Local officials had gradually turned to more lucrative activities such as industrial development, and the interest of the central state in agriculture had known a cruel drop. However, the situation considerably changed at the beginning of the twenty-first century, when agriculture, for a number of reasons we are about to explore, was put back on the agenda of the central government.

2.3   AGRICULTURE BACK ON THE CENTRAL AGENDA

No other civilization has had such a continuous tradition of thinking about famine, and no other nation's modern history has been so influenced by hunger and famine. (Lilian Li, *Fighting Famine in North China*)

China is currently experiencing an urbanization process of which the scale and pace are unprecedented. In the course of the four decades that followed the opening up of the Chinese economy, a flow of several hundred million people migrated from the countryside to urban areas. Between 2000 and 2009, urbanization accelerated—with 15 to 20 million people going to cities each year—and the proportion of urban people rose from 36.2 percent to 46.6 percent (China Development Research Foundation 2013: 14). In November 2010, the sixth national census revealed that the urbanization rate had already reached the one that was forecasted for 2020. In 2011, for the first time in China's millennium history, the number of urban citizens outreached the number of rural dwellers, with 680 million people living in cities and 270 million people living in urban agglomerations of more 1 million people (World Bank Database).

Chinese rural dwellers migrating to cities are largely incited to do so by rural–urban inequalities, both in terms of revenue and infrastructures. Development policies that were conducted in rural areas in the 1980s and 1990s had a tremendous impact on poverty alleviation. However, despite the rise in government expenditures allocated to rural areas, their share in the national budget decreased, in favor of investments allocated to urban areas and the industrial sector. The ratio between urban and rural revenues widened, jumping from 1.71 in 1984 to 2.55 in 1994 and to 3.2 in 2003–2004 (Fig. 2.1). Inequalities also developed in terms of infrastructures. While cities invested in communication and transportation infrastructures and water and electricity networks, rural areas were lagging behind.[10]

The widening gap, both in terms of revenue and equipment, constituted—and still is—one of the main drivers of the rural exodus. The trend of urbanization shall keep its pace in the coming years. If the urbanization rate indeed reaches 75 percent in 2050 as forecasted by experts and international organizations, the country will have experimented, in just over 50 years, a transition that developed countries underwent over more than one century—at much different scales.

The analysis of urbanization in developed and developing countries shows that the process usually comes hand in hand with economic development (World Bank 2009: 58–59). The concentration of people in urban areas is indeed likely to have positive effects on economies of scale and on economic activities and consumption, by bringing people closer to markets and to a wider diversity of products. In the mind of Chinese officials, urbanization has become strongly associated with development and economic catching, and, as such, is highly encouraged by the government. As Chen Yuan, Chairman of the Board of Directors of China Development Bank, states it: "'Urbanization' symbolizes how civilizations progress in

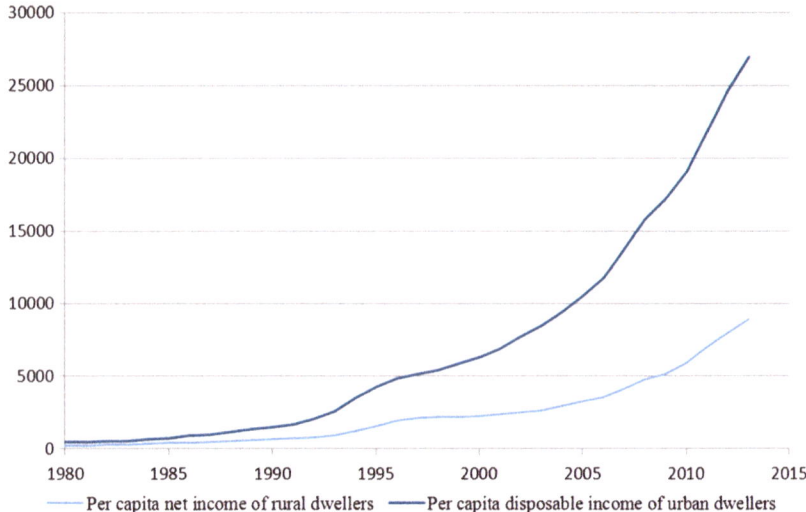

Fig. 2.1 The unequal rise of income of rural and urban dwellers (Source: Data from the National Bureau of Statistics of China)

general, but it also serves as the concentrated expression of a country's overall strength and international competitiveness" (China Development Research Foundation 2013: xix).

However, urbanization also brings a number of threats along, especially regarding food security and agriculture. The first consequence of China's rapid urbanization is a change in food diet. This evolution is not unique to China and other countries have experienced similar developments (Delisle 1990: 5). Higher income usually encourage people to diversify their food diet, and people living in urban areas have physical access to a greater variety of products compared to rural dwellers, for whom it is more difficult to go to stores, have access to processed food or use refrigeration. The revenue of Chinese urban dwellers being three times higher than the one of rural inhabitants, urban residents can afford buying more expensive products, such as meat, milk, and dairy products, and buying less grain. In the past decade, the rise in pork meat and milk consumption was the most pronounced rise and the demand for poultry and eggs also increased rapidly. Rural areas, helped by economic and infrastructure development, are following the same trends of evolution of food diets but are still far from catching up with consumption levels currently observed in urban areas.

In order to answer this rising food demand, the country needs to increase agricultural production. However, natural resources on which agricultural production relies on are highly limited. Water resources are scarce: the average renewable internal freshwater per capita was about 2062 cubic meters in 2014, or only one-third of the world average (World Bank Database). Water resources are also unevenly distributed across the territory: water availability falls to 500 cubic meters per capita in Northern China, whereas the south of the country is regularly affected by floods. Climate change is aggravating inter-regional differences: rainfall has been gradually declining in northern China (− 20 to − 40 mm per decade) and rising in the South of the country (+ 20 to 60 mm per decade) (Xie et al. 2009: 11). In addition, the melting of Himalayan glaciers, which feed the Huang and Yangzi rivers, the backbone of Chinese water resources, causes sudden floods, followed by worrisome periods of drought. Today, the annual water deficit would have reached 40 billion cubic meters (Zhang et al. 2009: 36). The situation should keep on following the same trend and water availability per capita might fall to 1890 cubic meters per year in the 2030s (Frenken 2011: 232).

Industrialization and urbanization aggravate water scarcity. Whereas Chinese agriculture used to be the main water consumer at the beginning of the 2000s, industrial and residential demands have been increasing rapidly over the last decade. In 2010, the share of agricultural demand has dropped to 61 percent, while the one of industry had gone from 13 percent up to 24 percent. According to some forecasts, urban water consumption could double by 2025 (Woetzel et al. 2009). The share of agriculture should keep on shrinking, while industrial and residential parts should reach 32 percent and 16 percent respectively by 2030 (Addams et al. 2009: 9). In addition, accelerated urbanization and industrialization led to major pollution issues. In 2006, according to a report, more than two thirds of the seven main Chinese rivers were unfit for human consumption (even after treatment), and almost one-third of their resources were completely useless, even for industrial or agricultural activities (Xie et al. 2009: 14). At the same time, agriculture, in China, highly relies on irrigation, as in the 2000s, 75 percent of grain was cultivated on irrigated land (Bruins and Bu 2006: 115).

The agricultural sector is not blameless regarding the degradation of water resources. The consumption of pesticides and fertilizers, highly encouraged by the government since the beginning of the 1980s, led to important problems of nonpoint source pollution.[11] Chinese farmers, in average, were using 565 kg of fertilizers per hectare in 2014. By comparison, farmers in the United States were using 138 kg per hectare (World Bank Database). Even if one takes into account the fact that there are several yields a year, especially in the South of the country, the gap is still huge. In addition, the imperfections of the subsidy system, the lack of training of farmers, and the reliance on potash imports led to imbalances in the use of fertilizers. Farmers generally consume too much nitrate fertilizers, at the expense of a balanced use of NPK.[12] The over-consumption and imbalances in the use of agricultural inputs prevent the soil from absorbing nutrients. Nitrate fertilizers, which are particularly subject to leaching, can contaminate groundwater wells that serve the cities. High levels of nitrate in water have adverse effects on human health and can also have disastrous consequences on aquatic ecosystems.

A lot of pressure is also exerted on arable land. The expansion of cities considerably erodes land available for farming. According to the estimates of the China Development Research Foundation (2013: 82), farmland areas dropped from 128 million hectares in 2000 to less than 122 million hectares in 2008, while space used for urban construction had risen by 36 percent in the same amount of time. In order to prevent a further diminution of the cultivated land, at the 11th People's Congress in March 2008, Yun Xiaosu, then vice-minister of Territory and Resources, set a "red line" of 1.8 billion mou of arable land.[13] However, the lucrative profits gained from the sale of urban land to real estate developers can contribute very significantly to the revenue of local governments. Land sales in Chengdu, for instance, would have accounted for 39 percent of the total revenue of the local government in 2005 (Woetzel et al. 2009: 87). In addition, as agricultural taxes were abolished in 2006, farming does not provide local governments with fiscal revenue as it used to do in the past. The fact that other economic sectors such as industry and trade are still taxed is another factor pushing local officials to favor the development of these sectors on cultivated land, at the expense of farming. Granting entrepreneurs with land also enables these latest to launch economic activities that generally contribute much more to local economic growth than agriculture. In a context where economic growth remains one of the most important evaluation criteria for local officials, it is quite easy to understand the rationale of land sales.

Land grabbing has become a matter of deep concern for the central authorities, as it seriously started threatening social stability in rural areas. The cause recently gained the support of the urban population, particularly active on social networks. The land of peasants has become a "legitimate right" (合法权利, *hefa quanli*), for which they are allowed to submit petitions. The defense of this legitimate right also enjoys the support of the central government—which is actively trying to curb the issue—starting an arm-twisting game with local governments. According to a number of analyses, land grabbing—although a widely spread practice throughout the country[14]—would not have had tremendous effects on the total arable land surface yet.[15] However, experts generally agree that there was a sharp decrease in the quality of arable land over the past few years. Arable land of the best quality is indeed usually located in the outskirts of cities, as historically, cities generally settled on areas with productive arable land able to feed the population. In the course of the growth of cities, the land located in the outskirts of settlements is the first to be converted into urban land, and arable land accounts for 57 percent of the area used by the recent expansion of cities in China (China Development Research Foundation 2013: 93). In order to keep figures intact, local governments often convert remote areas into arable land, whether they are suitable for agriculture or, on the opposite, located in arid, wet or mountainous areas.

Industrialization further aggravates the degradation of land. The eagerness of local officials to develop industrial activities in rural areas led to a lack of control of flue gas emissions and wastewater discharge. The accumulation of cadmium in rice crops is perhaps one of the most famous examples of industrial pollution in rural areas, which is regularly denounced by the Chinese media (e.g., in 2011, when an article of the Caixin New Century (Gong 2011) denounced that 10 percent of the rice sold on markets contained excessive rates of cadmium, or again in 2013 (Zhang 2013; Zheng 2013; Li 2013; Zheng and Gong 2013)). Over the past few years, the government took ever-stricter measures to regulate industrial flue gas emission and wastewater treatment. However, the local bureaus of the Ministry of Environmental Protection, which was granted the ministerial level only in 2008, still lack power as well as financial and human resources to effectively enforce regulations. In addition, local environmental protection bureaus also have to bargain with local cadres, among whom many are constrained by economic growth targets.

In addition, forest cover's losses, water diversions, over-exploitation of water resources, and changes in temperature caused by climate change damaged surfaces and led to serious erosion and desertification issues, even though the government started dedicating important efforts to forest conservation over the past few years (among others, through the

National Program for Forest Protection and the "Grain to Green Program", aimed at converting grain-sown areas into forests).

The degradation of land and water resources is worsened by temporary rural–urban migrations as well. Rural dwellers seeking to increase their income by working in cities off agricultural peak seasons have less time to work in fields. It encourages them to spread important volumes of agricultural inputs in fewer times (when they are in the countryside and available for farming activities), which worsens the efficiency of soil absorption, aggravates leaching, and further degrades land.

The decrease and degradation of arable land and water resources constitute important threats to sustainable food production. According to some experts, without efficient policies aimed at addressing the rarefaction of resources, grain yields could fall drastically in the coming years (Xiong et al. 2009). In parallel, urbanization and the improvement of living conditions led to an evolution of food diets, which became richer in meat, dairy products, and cooking oil, driving a rise in the demand for animal feeding (mostly maize and soybeans) and oilseeds. The inability to answer the growing grain demand forced the country to raise imports over the past few years. The agricultural trade balance became negative in 2004 and the deficit kept on growing since then (see Fig. 2.2).

The rising cost of the agricultural trade deficit is theoretically easily compensated by China's high trade surplus, which kept on increasing in spite of the world economic crisis. In 2012, the balance of trade was above €181 billion, up by almost 60 percent from 2011, as exports to the United States

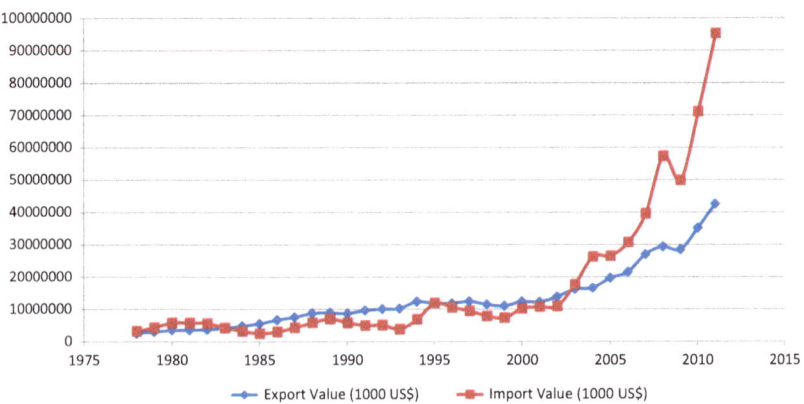

Fig. 2.2   Agricultural trade balance in China 1978–2011 (Source: FAO Database)

and Europe recovered (Directorate-General for Trade of the European Commission 2014: 9). However, the Chinese government attaches considerable importance to food security, establishing in 1996 a grain self-sufficiency target of 95 percent. Even if strong debates recently occurred for a revision of this target (questioning, in particular, the inclusion of soybean in the target (Chen 2011; Zhang 2012) and leading to the establishment of new targets by the prime Minister at the 2013 People's Congress of 90% for rice, wheat, and maize in the short and middle term (by 2015), and 80% in the long term (by 2025)), grain self-sufficiency objectives are still forming one of the most important guidelines of agricultural policies.

In the current context of globalization, the fear of famines inherited from the past can only partially explain the importance attached by present leaders to food security—even if Li (2007: 2) states that "no other civilization has had such a continuous tradition of thinking about famine, and no other nation's modern history has been so influenced by hunger and famine". During interviews, researchers working closely with the central government and central cadres mentioned two reasons to explain the government's willingness to maintain a high rate of food self-sufficiency. The first reason was "realism". I was explained that it was simply impossible for China to adopt a food strategy relying on imports like Japan (which imports about 60% of its food demand), given the demographic weight of the country. As was stating an expert from the Development Research Centre of the State Council and official of the Ministry of Agriculture:

> Food security is our number one goal. We need to support agriculture. [...] It is unlikely that China will follow the examples of Japan, which is relying on imports for 80 percent for its food demand, or Korea, which is 27 percent self-sufficient. [...] We cannot rely on international trade.[16]

The view of officials matches the view of a number of experts, according to whom, even if China would import just a small amount of its food demand, it would considerable destabilize global markets. Ni (2013: 5), for instance, states that "If China imports 10 percent of its current [cereal] consumption, its import volume will represent 20 percent of global imports".

The other reason mentioned by the interviewees was a willingness to guard the country against international price fluctuations. The price of products on international markets is not only about demand and supply but also a matter of currency exchange, which raises the possibility of price fluctuations.[17] The food price crises of 2007–2008, which saw the cereal price index reach a peak 2.8 times higher than in 2000, demonstrated the tremendous effects they could have on importing countries. Concerns also exist, among Chinese leaders, that food could be used as a weapon by foreign powers (Peng 2013).

Food security, especially in staple products such as grain, remains one of the most important goals of China's current agricultural modernization policies and was mentioned by all the interviewees from the central level.

The evolution of the stakes at hand in terms of food security and the willingness to improve social stability and economic development in rural areas have been putting pressure on the government over the last decade. Premises of the official re-emergence of agriculture and rural areas in the top-priorities of the government appeared at the end of the 1990s. According to Li (2007), at the beginning, the willingness of the state to engage in new reforms was not motivated by rising issues in rural areas, but rather by a willingness to reduce the power of local officials. The reform comprised two phases: during the first phase—from the release of the original reform package in 2000 to its implementation in 2003—the fiscal burden was reduced and the rural tax regime was rationalized. Many items were abolished, but agricultural taxes were raised in order to compensate townships for the losses generated in their income.

Farmers started benefitting from the reform only during its second phase of implementation. In 2004, the Number One Document—the first document issued by the State Council and the Central Committee of the Communist Party at the beginning of each year, which generally sets the tone of the policies that are to be promulgated throughout the year—introduced the concept of the *san nong* (三农), or the three rural issues:

> Under the guidance of the sixteenth Communist Party's National Congress, in 2003, various regions and departments, in accordance with the requirements of the central authorities, strengthened their will to solve the '*san nong*' issue, by withstanding the serious assaults of sudden outbreaks of SRAS, surmounting the high impacts of natural disasters that frequently occur, achieving the adjustment of the agricultural structure, steadily developing rural economy, deepening rural reforms, raising peasants' revenue and preserving and stabilizing rural society.

The 2004 Number One Document recognizes the issue of farmers' living conditions, stressing that a rise in farmers' income is a necessary step to address economic and political issues:

In the long term, the fact that farmers' income cannot increase will not only affect the living standards of these latest: it will also have an impact on food production and on the supply of agricultural products; it will not only hinder the development of rural economy, but it will also restrict the growth of the national economy; it will not only affect social progress in rural areas, it will also prevent on achieving the goal of building a well-off society; it is not only an major economic problem, it is also an important political issue.

The document encourages ministries and local governments to support agriculture, particularly in major grain producing areas. It recommends promoting the development, modernization, and industrialization of the agricultural sector and food chain, in order to improve the quality and safety of food products. The document also stresses the need to diversify the income sources of rural dwellers, by, among others, promoting the development of rural secondary and tertiary industries. Finally, the text emphasizes the need to strengthen infrastructures in the countryside. The three kinds of policies (addressing agricultural production, rural areas and farmers) are presented as strongly embedded in each other: for instance, the document underlines that building infrastructures in rural areas will help developing agricultural activities, which will in turn lead to a rise in farmers' income.

Almost all of the Number One documents that were published between 2004 and 2015 promulgated agricultural and rural development policy guidelines, except from the 2011 document, which focused on water conservancy (see Table 2.1). The evolution of the role given to the agricultural sector between the first half and the second half of the 2000s appears clearly when comparing the Tenth and the Eleventh Five-Year Plans. In the Tenth Five-Year Plan (2001–2005), agriculture was depicted as one lever of development among others. Agricultural development was mentioned only in the second chapter, among a whole set of tools aimed at "strengthening the economic structure". In comparison, the Eleventh Five-Year plan (2006–2011) introduces agricultural development as a fundamental and fully-fledged objective and dedicates a whole chapter to the "building of the socialist countryside". This chapter appears in second position, just after the chapter introducing general guidelines and objectives. The plan emphasizes that agriculture is not only useful to develop rural areas but also a pillar for the other economic sectors and addresses social and political issues. In addition, there are much more occurrences of the word 农 (*nong*) in the Eleventh Five-Year Plan than in the Tenth

Table 2.1   Titles of Number One Documents (2004–2014)

Year	Focus/main theme or goal
2004	Raising farmers' income
2005	Improving the overall production capacity of agriculture
2006	Building a "new socialist countryside"
2007	Developing modern agriculture and promoting the construction of a new socialist countryside
2008	Strengthening the foundations of agriculture
2009	Achieving steady agricultural development and rise in farmers' income
2010	Realizing coordinated urban–rural development and further strengthening the foundations of agricultural and rural development
2011	Accelerating the development of water conservancy
2012	Speeding up scientific and technology innovation to ensure adequate supply of agricultural products
2013	Accelerating the modernization of agriculture and further enhancing the vitality of rural development
2014	Deepening rural reform and accelerating agricultural modernization
2015	Enlarging the reform and bringing forth new ideas to speed up agricultural modernization

Five-Year Plan, including in chapters dedicated on issues other than the building of the socialist countryside (see Table 2.2).

The Twelfth Five-Year Plan (2011–2015) goes in the same direction that was established by the Eleventh Five-Year Plan, as the second part of the Plan already focuses on the "acceleration of rural and agricultural development". In this plan, again, the word 农 is mentioned an impressive number of times (Table 2.3).

The greater emphasis given to the agricultural sector was not just "virtually" established by central policy guidelines promulgated through Five-Years Plans and Number One Documents. Public expenditures dedicated to *san nong* issues also expanded dramatically in the years following the promulgation of the first Number One Document on rural issues (see Fig. 2.3).

Even though at first, expenditures were mainly allocated to the improvement of rural infrastructures, the government progressively built a comprehensive system aimed at directly supporting agricultural production. Agricultural taxes were abolished in 2006, relieving farmers from what had long been designated as "the burden of peasants".

The support system itself consists in several kinds of subsidies. The ones dedicated to agricultural inputs, such as pesticides and fertilizers, represent

Table 2.2   Frequency of occurrence of the word 农 in the Tenth and Eleventh Five-Year Plans

Tenth Five-Year Plan 2001–2005		Eleventh Five-Year Plan 2006–2010	
Chapters and occurrences of the word 农			
Chapter 1 Guidelines and objectives	3	Chapter 1 Guidelines and objectives	16
Chapter 2 Economic structure Main goals: **Str**engthening the foundations of agriculture and promoting the development of rural economy; Optimizing industrial structure and enhancing China's international competitiveness; Developing the service sector; Etc.	95	Chapter 2 Building the socialist countryside	172
		Chapter 3 Moving forward the optimization of the industrial structure	8
		Chapter 4 Accelerating the development of the services industry	3
		Chapter 5 Promoting a coordinated regional development	17
Chapter 3 Technology, education and talent	5	Chapter 6 Building an environmentally-friendly society and saving natural resources	7
Chapter 4 Population, resources and environment	11	Chapter 7 Rejuvenating the country through science and education and empowering the country through people's talents	15
Chapter 5 Reform and opening up	0	Chapter 8 Deepening institutional reform	0
Chapter 6 People's lives	5	Chapter 9 Implementing the strategy of mutually beneficial opening up	0
Chapter 7 Intellectual civilization	0	Chapter 10 Moving forward the building of a harmonious socialist society	4
Chapter 8 Legal system	0	Chapter 11 Strengthening the building of socialist democratic politics	0
Chapter 9 National defense	0	Chapter 12 Consolidating the socialist culture	3
Chapter 10 Implementing the plan	0	Chapter 13 Strengthening the national defense	0
		Chapter 14 Implementing the program	11
TOTAL occurrences of the word 农 119			256

Table 2.3   Frequency of occurrence of the word 农 in the Twelfth Five-Year Plan

Twelfth Five-Year Plan 2011–2015	
Chapters and occurrences of the word 农	
Part I: Transform growth pattern and create a new scenario for scientific development	
Chapter 1: Develop the environment	3
Chapter 2: Guidelines	2
Chapter 3: Main objectives	4
Chapter 4: Policy guidance	9
Part II: Accelerate the building of a new socialist countryside	
Chapter 5: Accelerate the development of modern agriculture	38
Chapter 6: Expand the channels to increase rural income	33
Chapter 7: Improve rural production and living conditions	42
Chapter 8: Improve institutional mechanisms for rural development	23
Part III: Transform and raise the competitiveness of core industry	2
Part IV: Build an environment to extensively develop the service sector	7
Part V: Optimize the structure and promote coordinated regional development and "healthy" urbanization	26
Part VI: Green development: build a resource-saving and environment-friendly society	9
Part VII: Innovation-driven: Implement science and education strategy and the development of new talents to reinvigorate the country	5
Part VIII: Improve people's livelihood: establish and improve basic public service systems	11
Part IX: Strengthen and innovate in social management	1
Part X: Pass on innovation: Extensively promote prosperous cultural development	3
Part XI: Reform and improve the socialist market economic system	1
Part XII: Improve the level of opening-up for mutual benefit	3
Part XIII: Democratic development: Promote the establishment of a socialist political civilization	0
Part XIV: Deepen cooperation: Build a common homeland for the Chinese people	0
Part XV: Civil-military integration: Strengthen the construction of national defense and the modernization of the army	0
Part XVI: Strengthen implementation and coordination of the plan	2
Total occurrences of the word 农	224

the largest share of expenditures. Subsidies are not always granted to farmers and can instead benefit input producers. Since 2003–2004, fertilizer producers, for instance, enjoy preferential prices for electricity, gas, coal, or transport. In addition, they are also granted abatements of VAT and of export taxes for the export of finished products. Finally, producers can have access to preferential loans for the building of production and storage

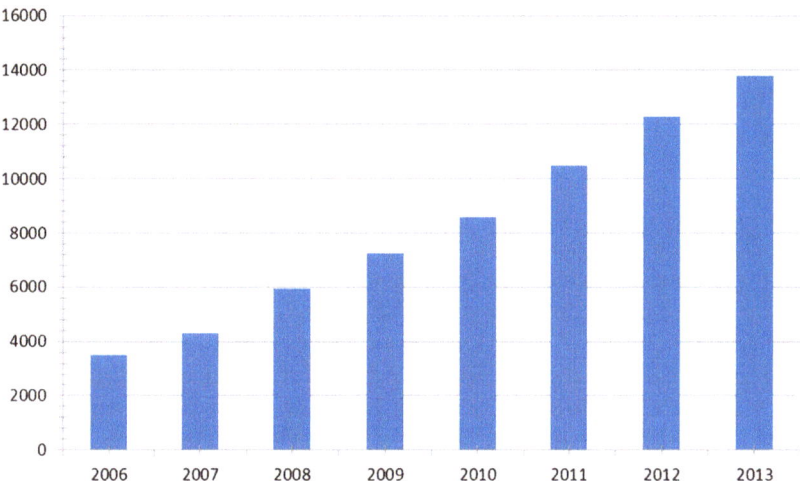

Fig. 2.3 *San nong* public expenditures (100 million RMB) (Source: 财政部,财政支持"三农"情况 *Caizhengbu, caizheng zhichi "sannong" qingkuang* [Ministry of Finance, Financial support situation for the three rural issues] http://www.mof.gov.cn/zhuantihuigu/czjbqk1/czzc/201405/t20140507_1076149.html)

infrastructures. All these mechanisms lower production costs and producers are then supposed to pass on price reductions to the farming sector. Subsidies for agricultural inputs also include subsidies for the purchasing of improved seeds. Financial support was first established for soybeans in 2002, before spreading to rice, wheat, and maize in 2004 and 2005. Allocation systems are within the jurisdiction of each province and depend of the size of their financial reserves for grain programs. The amounts of the subsidies as well as the allocation methods differ greatly from one region to another. Year 2002 also saw the development of subsidies for agricultural machinery. Financial support targets farmers, cooperatives, and/or producers, according to rules set by local governments.

Direct payments have only recently been introduced in the subsidy scheme and they still constitute a small part of the whole scheme. Their amounts differ greatly from province to province. Originally introduced with the aim of compensating grain growers for the rise in the price of agricultural inputs, they target mostly grain-producing areas. The subsidy scheme is finally completed by procurement and storage policies, which essentially target grain—even if the state can also intervene in markets through the purchasing or selling of other commodities such as pork. Grain is bought by the three state-owned enterprises (SOEs), Sinograin, COFCO, and China Tex, at minimum market prices annually set by the NDRC.

2.4   Conclusion

This chapter demonstrates how crucial the roles played by peasants, rural areas, and agriculture were in state building at the dawn of the PRC. The fertile ground for revolution of exploited peasants was turned into a real symbol by the nascent Communist Party, who took land reform as a rallying cry to spread revolution throughout the country. Even after their leadership was established, communist leaders kept on relying on rural areas for state-building, through the establishment of a nation-wide agricultural project with the development of the People's Communes.

In the 1980s, after the abolition of the People's Communes and the implementation of the HRS, the state progressively lost its interest in agriculture, focusing more on urbanization and industrialization. Several institutional reforms further weakened the involvement of governmental actors in agricultural production activities, such as the progressive abolition of procurement schemes, the declining importance of state planning in the agricultural sector, and the instauration of village elections.

However, on the eve of the twenty-first century, issues linked to agriculture and rural areas started to seriously worsen. Food security concerns resurfaced when the agricultural trade balance of China became negative in 2004, following an evolution of food diets and the degradation of production resources. In addition, rural–urban gaps have been keeping on widening, posing social stability threats. Faced to the necessity to address these issues, the government, in the middle of the 2000s, started promulgating policy guidelines focusing on rural and agricultural modernization and increasing dedicating budgets. Are these concerns transmitted down to local areas? Which strategies are implemented by central and local states to encourage agricultural modernization? The next chapters aim at examining a number of elements addressing these questions.

Notes

1. One catty = 0.5 kilogram.
2. One *tan* = 0.05 metric ton.
3. In China, "grain" (*liangshi*) not only refers to cereals but also encompasses peas and tubers.

4. Original language: "坚持适应社会需求结构的变化和国民经济现代化的要求,进一步合理调整产业结构" *Jianchi zhiying shehui xuqiu jiegou de bianhua he guomin jingji xiandaihua de yaoqiu, jinyibu heli tiaozheng chanyejiegou.*

5. Original language: "坚持恰当地确定固定资产投资规模,合理调整投资结构,加快能源、交通、通信和原材料工业的建设" *Jianchi qiadang dangdi queding guding zichan touzi guimo, heli tiaozheng touzi jiegou, jiakuai nengyuan, jiaotong, tongxin he yuancailiao gongye de jianshe.*

6. The adverse impacts that the newly established dual grain markets had on grain production, grain imports, and inflation pushed the government to reaffirm its role on the market in 1994. The complete liberalization of grain markets did not occur before the beginning of the 2000s. Reforms were conducted according to a progressive scheme: new grain market reforms were first implemented in 2001, before the State Council issued new regulations that completely liberalized markets in 2004.

7. It will be raised at 30 years in the 1990s.

8. From an average growth rate of 4.9 percent between 1970 and 1978, the rate of increase jumps to 8.8 percent in average between 1979 and 1984, before reaching a peak of 15 percent on 1984 (Tongeren and Huang 2004: 27–28).

9. In certain cases, the leader of the village committee and the Party secretary are one and the same person. In villages I visited, villagers explained that because of urban–rural migration, few people capable of taking over political functions remained in the village. It was the reason given for the overlap of responsibilities between the Party secretary and the supposedly independent village committee.

10. Many places I went to in the countryside did not have flushing toilets, and in a number of areas, households still had to use wells for water supply. The education system was usually considered as poor by rural dwellers in most places. Roads linking small villages to towns had just recently been paved.

11. Nonpoint source pollution refers to water and air pollution from diffuse sources. One major source of water pollution from diffuse sources is due to the runoff of chemical fertilizer into the soil and underground water.

12. N: nitrogen; P: phosphate; and K: potassium are the three main macronutrients of chemical fertilizers, that enhance the growth of plants. The equilibrium—the proportion of N, P, and K farmers should apply on their soil—depends on several elements such as the type of crop, the state of growth, the physical properties of the soil, and climate conditions.

13. 120 million hectares.

14. Between 2008 and 2011, land inspectors found 64,366 cases of land violation, involving more than 240,000 hectares (Yu 2011).

15. According to official data, China's arable land would have shrunk from 130 million hectares to 122 million hectares from 1996 to 2004 (http://www.gov.cn/english/2005-10/24/content_82778.htm accessed on 28 January 2014), and remained above 120 million hectares since. Data on arable land though differ widely from one source to another. According to FAO's estimates, China's arable land would have dropped from 119,339 million hectares in 1996 to 105,920 million hectares in 2009 (FAO Database).
16. Closed-door conference, Beijing, October 2012.
17. Presentation of the Director of the Department of Rural Economics Research of the Development Research Center of the State Council, closed-door conference, Beijing, October 2012.

REFERENCES

Addams, L., Boccaletti, G., Kerlin, M., Stuchtey, M., et al. (2009). *Charting our water future*. Washington, DC: 2030 Water Resources Group.

Ahn, C. S. (1997). Government-party coordination in Japan's foreign policy-making: The issue of permanent membership in the UNSC. *Asian Survey, 37*(4), 368–382. https://doi.org/10.2307/2645654.

Aird, J. S. (1982). Population studies and population policy in China. *Population and Development Review, 8*(2), 85–97.

Ash, R. F. (1992). The agricultural sector in China: Performance and policy dilemmas during the 1990s. *The China Quarterly, 131*, 545–576. https://doi.org/10.1017/S0305741000046294.

Bouvier, C. (1958). *La collectivisation de l'agriculture: URSS – Chine – Démocraties populaires*. Paris: Armand Colin.

Bowie, R. R., & Fairbank, J. K. (1962). *Communist China 1955–1959: Policy documents with analysis*. Cambridge: Harvard University Press.

Bruins, H., & Bu, F. (2006). Food security in China and contingency planning: The significance of grain reserves. *Journal of Contingencies and Crisis Management, 14*(3), 114–124. https://doi.org/10.1111/j.1468-5973.2006.00488.x.

Chen, T. (1974). *The Maoist educational revolution*. New York: Praeger.

Chen, Z. (2008). Urbanization and spatial structure: Evolution of urban system in China. Institute of Developing Economies, Japan External Trade Organization. Visiting Research Fellows Series 439.

Chen, J. (2011, August 9). Grain imports and China's food security. *Diaoyan Shijie* (56) [陈洁, 粮食进口与我国的粮食安全, 调研世界 Chen Jie, Liangshi jinkou yu zhongguo de liangshi anquan. *Diaoyan Shijie*].

China Development Research Foundation. (2013). *China's new urbanization strategy*. Abingdon/New York: Routledge.

China Water Risk. http://chinawaterrisk.org/big-picture/2030-demand-supply/. Accessed 16 July 2014.

Delisle, H. (1990). *Patterns of urban food consumption in developing countries: Perspective from the 1980's.* Food Policy and Nutrition Division FAO, Rome ftp://ftp.fao.org/es/esn/nutrition/urban/delisle_paper.pdf

Dikötter, F. (2010). *Mao's great famine: The history of China's most devastating catastrophe, 1958–62.* London: Bloomsbury.

Directorate-General for Trade of the European Commission. (2014, January 16). Trade European Union, trade in goods with China. *Bruxelles.* http://trade.ec.europa.eu/doclib/docs/2006/september/tradoc_113366.pdf. Accessed 20 July 2014.

Feng, Y., He, D., & Beth, K. (2006). Water resources administration institution in China. *Water Policy, 8,* 291–301.

Frenken, K. (2011). *Irrigation in Southern and Eastern Asia in figures.* AQUASTAT Survey. Rome: FAO Land and Water Division.

Gong, J. (2011). The thoughts of killing of cadmium rice. *Caixin New Century* (6) [宫靖, 镉米杀机. 财新《新世纪》, 2011年第6期 Gong Jing, ge mi shaji. *Caixin 'xin shiji'*].

Greer, A. (1999). Policy coordination and the British administrative system: Evidence from the BSE enquiry. *Parliamentary Affairs, 52*(4), 598–615.

Harrison, P. (1972). *The long march to power: A history of the Chinese Communist Party, 1921–72.* London: Macmillan.

Huang, J., & Rozelle, S. (1997). Technological change: Rediscovery of the engine of productivity growth. *China's Rural Economy Journal of Development Economics, 49*(2), 337–359. https://doi.org/10.1016/0304-3878(95)00065-8.

Huang, J., & Rozelle, S. (2009). Développement agricole et nutrition: Politiques à l'origine du succès chinois. Document Hors Série n°19, Programme Alimentaire mondial.

Kerkvliet, B., Chan, A., & Unger, J. (1998). Comparing the Chinese and Vietnamese reforms: An introduction. *The China Journal, 40,* 1–7. https://doi.org/10.2307/2667451.

Li, L. (2007). *Fighting famine in north China: State, market and environmental decline, 1690s–1990s.* Stanford: Stanford University Press.

Li, H. (2009). *Village China under socialism and reform: A micro history, 1948–2008.* Stanford: Stanford University Press.

Li, X. (2013). Tracing to the source the rice controversy in Hunan and Guangdong. *Caixin New Century* (20) [李雪娜. 湘粤米争溯源. 财新《新世纪》, 2013年第20期 Li Xuena, xiang yue mi zheng suyuan. *Caixin "xin shiji"*].

Li, X., Wang, D., Jin, L., & Zuo, T. (2006). *Impacts of China's agricultural policies on payment for watershed services.* London/Beijing: College of Humanities and Development/China Agricultural University and International Institute for Environment and Development.

Lieberthal, K. G. (1992). Introduction: The "fragmented authoritarianism" model and its limitations. In K. G. Lieberthal & L. GDM (Eds.), *Bureaucracy, politics and decision-making in post-Mao China*. Berkeley: University of California Press.

National Bureau of Statistics Database. http://data.stats.gov.cn/workspace/index?m=hgnd

Ngo, T. M. (2009). A hybrid revolutionary process: The Chinese cooperative movement in Xiyang County, Shanxi. *Modern China, 35*(3), 284–312. https://doi.org/10.1177/0097700408328972.

Ni, H. (2013). *Agricultural domestic support and sustainable development in China*. ICTSD Programme on Agricultural Trade and Sustainable Development, Geneva.

Østergaard, C. S. (1990). Introduction. In J. Delman, C. S. Østergaard, & F. Christiansen (Eds.), *Remaking peasant China: Problems of rural development and institutions at the start of the 1990s*. Aarhus: Aarhus University Press.

Peng, G. (2013, August 21). 8 expert's questions on GM-staple grain: Why exactly does our country import unnecessary things? *Huanqiu Journal*. [彭光谦, 专家八问主粮转基因化:我国究竟为何要盲目引进，环球时报，2013年08月21日 Peng Guangqian, Zhuanjia ba wen zhuliang zhuanjiyinhua: wo guo jiujing weihe yao mangmu yinjin. *Huanqiu shibao*] http://news.xinhuanet.com/food/2013-08/21/c_125216637.htm. Accessed Sep 2013.

Sun, S. (2006). *The long march*. London: Harper Collins.

Tongeren, F. W., & Huang, J. (Eds.). (2004). China's food economy in the early 21st century; development of China's food economy and its impact on global trade and on the EU. The Hague: Agricultural Economics Research Institute (LEI).

Vermeer, E. B. (1998). Industrial pollution in China and remedial policies. *The China Quarterly, 156*, 952–985.

Waldron, S., Brown, C., & Longworth, J. (2006). State sector reform and agriculture in China. *The China Quarterly, 186*, 277–294. https://doi.org/10.1017/S030574100600015.

Wilson, D. (1971). *The Long March 1935: The epic of Chinese communism's survival*. London: H. Hamilton.

Woetzel, J., Mendonca, L., Devan, J., Negri, S., Hu, Y., Jordan, L., Li, X., Maasry, A., Tsen, G., Yu, F., et al. (2009). *Preparing for China's urban billion*. Washington, DC: McKinsey Global Institute.

World Bank. (2009). *World development report*. Washington, DC: World Bank.

World Bank Database. http://data.worldbank.org/

Xie, J., Liebenthal, A., Warford, J. J., Dixon, J. A., Wang, M., Gao, S., Wang, S., Jiang, Y., & Ma, Z. (2009). *Addressing China's water scarcity recommendations for selected water resource management issue*. The International Bank for Reconstruction and Development/The World Bank.

Xiong, W., Conway, D., Lin, E., Xu, Y., Ju, H., Jiang, J., Holman, I., & Li, Y. (2009). Future cereal production in China: The interaction of climate change, water availability and socio-economic scenarios. *Global Environmental Change, 19*(1), 34–44. https://doi.org/10.1016/j.gloenvcha.2008.10.006.

Yang, B. (1990). *From revolution to politics; Chinese communists on the long march.* Boulder: Westview.

Yu, H. (2008). Global governance against global warming and China's response: An empirical study on climate change policy coordination in China from 1992 to 2002. *Current Politics and Economics of Asia, 17*, 267–295.

于猛, 评估报告称地方政府"土地依赖症"成体制性障碍, 人民日报, 2011年11月25日 Yu Meng, pinggu baogao cheng defang zhengfu 'tudi yilai zheng' cheng tizhi xing zhang'ai, renmin ribao [YU, Meng. The local governments' sickness of land dependency is turning into an institutional obstacle. People's Daily, 25 November 2011]. http://politics.people.com.cn/GB/70731/16387913. html. Accessed 28 Jan 2014.

Yu, X., & Zhao, G. (2009). Chinese agricultural development in 30 years: A literature review. *Frontiers of Economics in China, 4*(4), 633–648.

Zhang, X. (2012, September 27). Chinese food security: Problem and measures. *Aisixiang.* [张晓山, 中国的粮食安全问题及其对策, 爱思想, Zhang Xiaoshan, Zhongguo de liangshianquan wenti jiqi duice. *Aisixiang*] http://www.aisixiang.com/data/57747.html

Zhang, C. (2013, July 15). Where does the illness of legal proceedings for environmental public interests stand? *China Dialogue.* [张 春. 环境公益诉讼 病在哪里?中外对话, 2013年17月15日 Zhang Chun, huanjing gongyi susong bing zai nail? *Zhongwai duihua*] https://www.chinadialogue.net/article/show/single/ch/6206-Cadmium-pollution-in-Yunnan-reopens-debate-over-public-interest-litigation-

Zhang, J., Wang, G., Yang, Y., He, R., & Liu, J. (2009). Impact of climate change on water security in China. *Advances in Climate Change Research, 5*, 34–40.

Zheng, D. (2013). Save rice. Caixin New Century (20) [郑道. 拯救大米. 财新《新世纪》, 2013年第20期 Zheng Dao, zhengjiu dami. *Caixin "xin shiji"*].

Zheng, D., & Gong, J. (2013). The cadmium rice controversy. Caixin New Century (20) [郑道, 宫靖. 大米镉超标争议, 财新《新世纪》, 2013年第20期 Zheng Dao, & Gong Jing, Dami ge chaobiao zhengyi. *Caixin "xin shiji"*].

CHAPTER 3

Enterprises: The New Leaders of Agricultural Modernization

3.1 Rural Food-Processing Enterprises to Address Price and Food Safety Issues

3.1.1 Enterprises as Historical Players in Rural Areas

The choice of relying on enterprises to conduct economic development is not new, especially in China, and especially in rural areas. From the encouragement of the multiplication of Township and Rural Enterprises by local governments in the 1980s to the state-led privatization of collectively owned enterprises and to the controlled development of private enterprises in the 1990s and 2000s, rural enterprises have always played a major role in rural economic development.

In the 1980s, the decrease in the importance of state economic planning and the reforms of the fiscal system granted local governments with new powers—which was not without posing problems (Oi et al. 2012). In the aftermath of fiscal decentralization, local officials, "fiscally autonomous" for a certain number of items, had to face an increasing pressure to develop economic activities in their area of jurisdiction, especially since the abolition of the People's Communes had deprived township and village governments from major income sources. In addition, the new cadres evaluation system established economic growth achievements as one of the main evaluation criteria. As poor economic performance, from then

© The Author(s) 2018
M.-H. Schwoob, *Food Security and the Modernisation Pathway in China*, Critical Studies of the Asia-Pacific,
https://doi.org/10.1007/978-3-319-65702-8_3

on, could adversely influence the career of local officials, these latest became particularly eager to concentrate on economic development. Local leaders were not only exhorted to achieve economic growth by the new pressures put onto them by fiscal decentralization and by the reform of the cadres evaluation system. They were also given greater incentives to do so. Promoting industrial growth was indeed a way to get wealthier, thanks to the institutional settings of the cadres responsibility system, such as the establishment of direct links between the income of local cadres and the local industrial performance (Whiting 2000: 107).

The most important effect of these new incentives and pressures was perhaps the tremendous development of TVEs (乡镇企业 *xiangzhen qiye*). Many scholars consider that the development of TVEs was the main driver of economic growth in rural areas and highly contributed to the rise and diversification of revenues (De Janvry et al. 2005). Some even regard TVEs as one of the most important drivers of the national economic growth that was achieved in the 1980s and 1990s. As Wang (2005: 177) state it: "Throughout the reform period, township and village enterprises (TVEs) have constituted one of the most dynamic sectors in the Chinese economy." Township and Rural Enterprises developed tremendously in the 1980s. From only 1.4 million in 1980, their number rose to almost 19 million in 1988 (National Bureau of Statistics). One of the most important factors that contributed to this rapid development was the considerable agricultural labor surplus generated by decollectivization. Local cadres had to find a way to create new jobs for the rising number of unemployed people in rural areas, as peasants, weakened by the era of collectivization, were economically unable to create enterprises themselves. In the end, from 1978 to 1996, TVEs absorbed 110 million of laborers coming from the agricultural sector (Pei 2002: 282).

TVEs were collectively owned but had few things in common with the Maoist collective systems once the HRS was established. According to Pei (2002: 289), TVEs were then "relatively independent [from the state control] and community oriented." For the author, this difference played a significant role in the development of TVEs in rural areas. This does not mean that local governments were not involved in the process. They were, on the opposite, quite active players in the development of TVEs. In reality, the ownership structure of TVEs was not very different from the one of SOEs. Residents of villages and townships, supposed to be among the owners, were in fact represented by their village committee members or township government officials. For Oi (1999a: 66), the consequence was that "township and village enterprises allowed local officials to keep their control over the economy and to use this control to maintain their patron-client networks and personalized systems of authority".

At this time, suspicion vis-à-vis private enterprises persisted among local cadres, who were taking measures to hinder these latest to take off. TVEs, in this context, constituted a more comfortable solution to promote economic development without relying solely on SOEs. Public ownership of TVEs enabled local cadres to use traditional bureaucratic methods to control their management and operations. In addition, TVEs were usually run under a contracting scheme establishing output, profits, and revenue targets, such as in the era of state planning. To sum up, TVEs were answering economic development goals of local cadres without depriving them from their means of control over local economic stakeholders.

In the course of economic liberalization, private stakeholders progressively acquired the rights and conditions to develop business activities in rural areas. Individuals had enough time to accumulate capital to create enterprises. In addition, local officials, challenged by falling profits of rural activities and increasing deficits of TVEs, had gradually changed their minds about the threat posed by private enterprises and started consenting to their development. The same reasons progressively pushed local cadres to change the ownership structure of TVEs, which experienced reforms similar to the ones SOEs underwent in parallel (Han 2003). In most cases, local leaders initiated the privatization process (Li and Rozelle 2003: 991). Firms were sold to "insiders", such as managers and employees (Kung 1999). By the mid-1990s, local leaders had already privatized more than half a million collectively owned enterprises (Oi 1999b: 624). The active role local leaders played in the development of the capitalist sector created state-business nexus, through which local leaders could both promote and control the development of this burgeoning private sector.

The social and institutional ties linking local state agents with the managers of newly privatized firms or with private entrepreneurs were thoroughly investigated by an abundant research, and, in particular, by the proponents of Chinese corporatism. Whereas in its early stages, "corporatism" was referring either to a particular form of state involvement (through economic corporations) mainly promoted by the fascist ideology, or to the cooptation of trade unions by the state in the framework of labor movements, today, the most commonly used definition for what has since been retermed "neo-corporatism" is perhaps the one given by Schmitter (1974: 93–94), who describes it

as a system of interest representation in which the constituent units are organized into a limited number of singular, compulsory, noncompetitive, hier-
archically ordered and functionally differentiated categories, recognized or
licensed (if not created) by the state and granted a deliberate representa-
tional monopoly within their respective categories in exchange for observing
certain controls on their selection of leaders and articulation of demands
and supports.

After having acknowledged a similar "ideal-type" of corporatism,[1] Unger
and Chan (1996: 95) recognize that the concept has evolved and been
broadened to include various forms of corporatism, including the "soci-
etal corporatism" of democratic countries. The authors explore the characteristics of corporatism in several Asian countries, before taking the
example of China, for which they develop a particularly interesting theory of "state-corporatist model". Unger and Chan argue that the state
has been able to maintain control over the society and over the (privatizing) economy thanks to corporatist structures serving as bridging agents.
These corporatist structures, established prior to the reform era (e.g.,
industrial unions and peasants associations), play the role of "transmission belts [...] providing a two-way conduit between the Party center and
the assigned constituencies" (Unger and Chan 1996: 104). These
"proto-corporatist structures," which did not fulfill their role of percolat-
ing demands up to the central government during the Maoist era, began
to operate as real corporatist structures in the aftermath of the reform,
when the system started loosening up at the beginning of the 1980s. In
addition to former structures, the state authorized the registration of new
associations, which act as additional corporatist intermediaries and agents.

The model of "local state corporatism" developed by Oi differs slightly
from the description made by Under and Chan. By local state corporatism, Oi (1992: 100–101) refers to "the workings of a local government
that coordinates economic enterprises in its territory as if it were a diversified business corporation". In this model, local governments keep control
over enterprises through several means: Firstly, through the contract
responsibility system sometimes in operation, and under which the role of
dictating the disposition of enterprises' profits remains with local governments; secondly, through the allocation of key resources (whether it be
state-supplied goods such as steel and cement—of which the quantities are
limited—or goods that are scarce in rural areas, such as fuel, oil, electricity,
and raw materials); thirdly, through the providing of bureaucratic services,
such as help in securing licenses, certification and prizes for products, and

tax breaks; finally, through investment and credit (as in rural areas, enterprises need a guarantor to secure a loan, a role which can be taken by the township economic commission). Privatization, in rural areas, therefore did not put an end to state-enterprises nexus that had mushroomed under the era of TVEs' development. On the opposite, local states managed to establish new ties with private entrepreneurs as well as corporatist structures linking them with new economic circles and granting them with new instruments of promotion and control. As the development of private enterprises in rural areas was seen as a way to fulfill economic development goals that local cadres had to achieve, these latest have long recognized and valorized the role enterprises could play for a number of development issues. This generated an important path dependency that profoundly influences the current strategy implemented by local governments for agricultural modernization: to favor food-processing enterprises based in rural areas and use them as transmission belts.

Local officials could have chosen different strategies to modernize agriculture. For instance, they could have relied on noncorporatist strategies, as they used to do until recently. Following the establishment of the HRS in the 1980s, the government indeed let unorganized individual farmers take possession of the agricultural sector, and farmers have operated relatively independently from the grip of the Party-state since then. Local officials could also have relied on different corporatist strategies, for instance, by involving actors other than enterprises such as farmers' unions. For rural economic sectors others than agriculture, it is as if the development of corporatist methods favoring industrial entrepreneurs was a necessary consequence of economic liberalization: local officials, unable to retain the development of private enterprises any longer, had to establish close links with entrepreneurs (in rural areas, with industrial entrepreneurs, in particular) in order to keep political and economic control over local players. In the current context of agricultural modernization, the rationale of the choice of local officials to establish privileged relations with entrepreneurs of rural industries is more difficult to explain. How can the hundreds of millions of farmers, whom agricultural production depends on, be excluded from corporatist structures, described as one of the most important control mechanisms of the Chinese government over its economy? What is exactly the role played by enterprises in the course of agricultural modernization? The following analysis of the rising challenges of food-price inflation and food safety and of local patterns of power provides answers to these questions.

3.1.2   *"Melting Food Chains" to Curb the Issue of Food Prices*

In terms of food security, the Chinese government is faced to the need to address several issues. Among the four dimensions of food security (physical availability, economic and physical access to food, nutrition, and the stability of the three mentioned dimensions over time) (FAO 2008), economic access to food, in particular, has long retained the attention of the Chinese government. In China, food prices have a major impact on national inflation. Food commodities still constitute an important part of the average basket of goods and services. According to the National Bureau of Statistics, food expenditures still accounted for about 35 percent of urban and rural budgets in 2012 and could reach 43 percent for poor rural households.

However, addressing the issue of economic access to food through a limitation of the rise in the price of food commodities is likely to have adverse effects on the livelihood of food producers. This is not a problem unique to China. While many developing countries struggled with soaring food prices in 2007–2008 and in 2013, the 2014 drop in grain prices negatively affected the income of farmers. According to the FAO (2011: 8), around 75 percent of the world's one billion hungry people are small-scale farmers, fishers, and foresters, who "depend entirely on agriculture and related enterprises for their food security and livelihoods". Only about 20–25 percent of the world's hungry people live in urban area, even though the number of hungry rural dwellers is rising rapidly with worldwide urbanization.

In China, there are still about 350 million farmers, who increasingly have to cope with higher production costs due to the rise in the price of fuel, fertilizers, and labor. Curbing price increases, in addition to potentially fueling protests among farmers—who already suffer from low economic conditions—would only push more labor force out of the farming sector. In order to address these threats, the government established in the 2000s minimum price policies for wheat, rice, and corn, along with a national food procurement and storage system. Mostly targeting grain, the system aims at fulfilling three objectives: guaranteeing a minimum price to farmers, as a way to encourage these latest to keep on growing grain; guarding the country against major grain shortage, for instance, in the event of a natural disaster; protecting consumers from price increases of basic staple products. Minimum prices are set annually in November by a committee gathering officials from the MOA, from the State

Administration of Grain, the Ministry of Finance and Sinograin (the most important SOE in charge of grain storage in the country). An average price is determined according to several criteria, such as the minimum price established for the previous year, the evolution of production costs, the stock levels in major producing areas, and the expected levels of production. The "average price" is then adjusted according to the variety and the quality of the grain purchased by state granaries, to the period and to the purchase location—the program only targets a few producing provinces. Whenever the market price at the farm gate falls below the minimum price established by the central government, Sinograin, along with two other SOEs, China National Cereals, Oils and Foodstuffs Corporation (COFCO) and ChinaTex, are requested to buy grain from farmers at the minimum price or above. These massive purchasing programs are supposed to trigger a market response that makes the price of grain rise. In order to be able to purchase grain above market prices, Sinograin, COFCO, and ChinaTex benefit from loans of the Agricultural Development Bank as well as from governmental subsidies according to the number of silos they are able to fill with grain.

For the past three years, minimum prices established by the government for rice, wheat, and maize have grown by 15–20 percent annually, in order to enable farmers to cope with the rise in production costs. In addition, there has been an important appreciation of the Chinese currency against the US dollar. As a result, by 2011, the prices of most major agricultural commodities, in China, were already 20–30 percent higher than US prices (Gale 2013) and the price of rice, wheat, and maize has remained above international market prices since. One of the consequences is that it became way more interesting for Chinese millers and other food-processing companies to buy grain from abroad—even when domestic markets produce enough to answer the national demand. In addition, the storage system represents an increasingly unbearable burden for the government, which is currently implementing pilot projects to find alternative systems to support grain production, such as target prices or direct subsidies.

In order to address the issue of rising food prices without harming farmers, the Chinese government recently initiated another strategy, by targeting intermediaries in the food chain. In China, the food chain is characterized by a high number of intermediaries, who link urban markets with small, remote, and scattered farmers in the countryside. According to the National Bureau of Statistics, the average size of farms is less than one hectare.[2] These data should be taken with extreme caution. Firstly, data regarding the average size of farms are extremely complicated to collect in China, due to the remoteness of farmers, to the informality of rental markets and to the (temporary or permanent) mobility of the agricultural workforce. In addition, this figure hides considerable discrepancies between farms (some state farms can be more than 1000 hectares). Having taken into consideration these remarks, it can be asserted that the size of the vast majority of Chinese farms is still extremely small compared to European countries or to the United States.

The small size of farms poses several issues. As many small farmers still do not have vehicles suited for the transport of agricultural products to markets, intermediaries able to come to the farm gate to purchase agricultural products are necessary players to link producers with consumers. In addition, agricultural production is scattered among a wide number of small producers, sometimes located far from consumption centers. The distance between production and consumption sites multiplies the levels of intermediation, from the smallest wholesalers, who simply link farmers with small wholesale markets in township-level cities, to the largest ones who trade on wider and sometimes more diversified wholesale markets, able to answer the needs of modern urban supermarket chains, in terms of volumes and diversity of products. As was stating a sales manager of a supermarket chain in Shanghai:

> On our modern supply chain, we work a lot with brokers. The main interest of brokers is that they sell apples, but also nuts, vacuum cleaners. [...] Beside, they are able to trade very big volumes. (Interview, Shanghai, June 2012)

In addition, I was told that regulatory constraints sometimes prevented retailers from signing commercial contracts directly with processing plants, as these latest usually did not hold the appropriate business license (partly because of protectionist policies implemented by provinces), whereas brokers usually did. The lack of knowledge and professionalism of farmers and processing plants in terms of sales was another important factor mentioned by the interviewees, justifying the existence of a large number of brokers (Fig. 3.1).

The Chinese government recently started targeting the intermediaries of the food chain. "Melting" food chains by cutting intermediaries out would reduce the price of food for final consumers by deducting their margins, without negatively impacting farmers. However, the problem

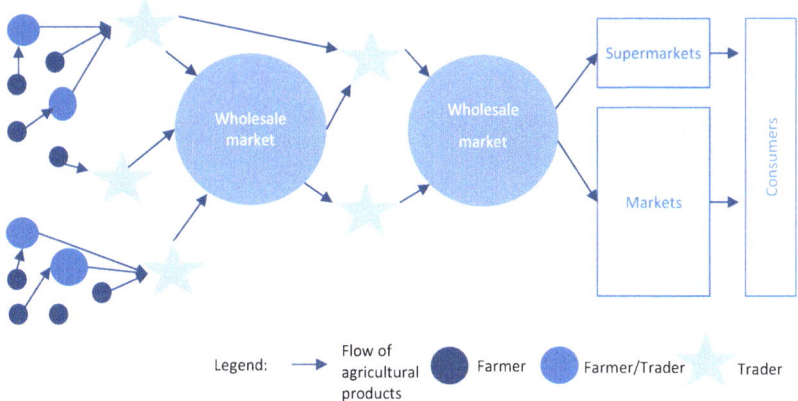

Fig. 3.1   Current Chinese food chain

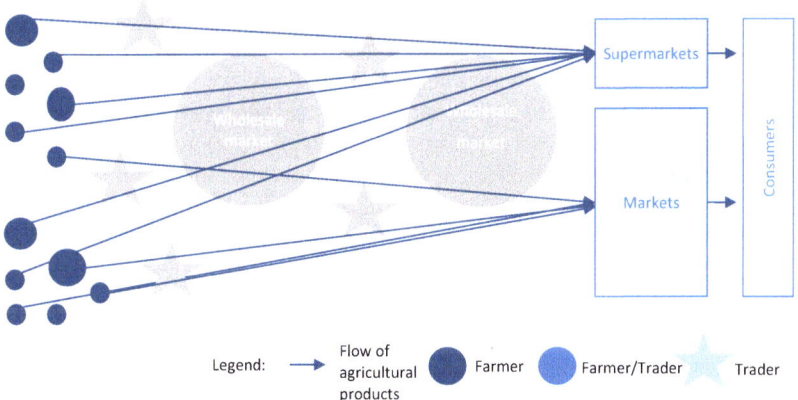

Fig. 3.2   Food chain without intermediaries

that farmers are usually smallholders scattered all over the countryside and producing small volumes in remote areas remains. Implementing a food chain model such as that presented in Fig. 3.2 thus does not appear very credible, as it would increase transaction costs in a tremendous way for retailers.

As a consequence, the model privileged by local governments is to give a greater role to food-processing enterprises based in rural areas (Fig. 3.3).

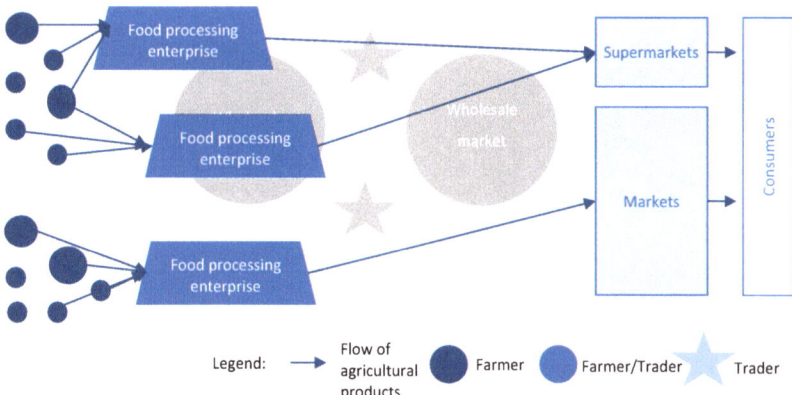

Fig. 3.3 Expected food chain

Insights from fieldwork in Jiangxi and Shandong led to several conclusions related to this matter. Firstly, even for simple fruits and vegetables, industrial processing is usually required to transform the initial product into a marketable product able to answer the needs of modern urban markets. For instance, apples produced in Shandong province have to be sorted, cleaned, waxed, and packed before being sold to retailers (Fig. 3.5). Similarly, oranges had to be sorted, cleaned, waxed, and packed (Fig. 3.4). Processing also often includes the artificial ripening of fruits—especially for oranges, which are more delicate products and have to be sold quickly once ripe.[3]

In addition, surveys conducted in rural areas demonstrated that food-processing enterprises were usually located in rural areas, close to farms or orchards. Comparatively to urban retailers, rural food enterprises suffer much less from transaction costs when establishing relationships with small producers located near to them, as they already operate with them on a regular basis for the supply of their processing plants. In the melting of food chains, rural-based food-processing enterprises become key intermediaries between farmers and urban retailers (Fig. 3.6).

3.1.3 Administration, Research and Civil Society Unable to Address Environmental Issues in the Agricultural Sector

In addition to rising food prices, the Chinese government is increasingly challenged by issues linked to the safety of food products, to which the

Fig. 3.4 Farmers-workers packing apples in Shandong (Photography by the author, Nov. 2012)

overuse of pesticides much contributes. During the last decades of the twentieth century, the direction taken by the government for agricultural development was the one of an input-intensive agriculture. The rationale behind this productivist view—which is not unique to China—is mainly based on the scarcity of land resources, as China has to feed almost 20 percent of the global population with only 7 percent of worldwide arable land. Even if this discourse has recently integrated other voices calling for a more rational use of pesticides, the current over-reliance on agricultural

Fig. 3.5 Farmers-workers sorting oranges in Jiangxi (Photography by the author, Oct. 2013)

inputs persists, partly because of path dependencies caused by past policies that have promoted the extensive use of agricultural inputs at the end of the twentieth century. In fact, the Chinese government was already implementing productivist agricultural policies under the collectivist era. However, these programs had mixed results at that time, given the dramatic situation in which the Chinese countryside was in the aftermath of the Great Leap Forward and given the tumultuous political, economic, and social context of the 1970s. During the Dengist era, reforms progressively helped farmers increase their consumption of agricultural inputs. The establishment of the HRS boosted their income, which enabled them to buy more pesticides, fertilizers, and improved seeds. In parallel, reforms were conducted in the industrial sector producing agricultural inputs. During the early stages of the reforms, the sector remained in the hands of the state. However, during the second half of the 1980s, input markets

Fig. 3.6 Workers unloading a truck coming back from farms located near an orange processing factory in Jiangxi (Photography by the author, Oct. 2013)

were progressively liberalized, starting with the ones of pesticides and agricultural machinery—the fertilizer market following at the end of the 1990s. While before reforms, the monopoly of the state enabled farmers to have access to cheap inputs, liberalization, on its side, significantly improved the supply of products, as it led to a rapid multiplication of producers in the countryside. Remote areas, in particular, which used to suffer from insufficient supplies of farm inputs, could from then on benefit from the technical advantages of the "Chinese green revolution." The multiplication of input producers was highly encouraged by the government, who started developing a comprehensive subsidy scheme targeting these actors. For Kung and Cai (2000: 278), there is no doubt that the increase in the use of chemical fertilizer, in the 1980s and 1990s, was "by and large a rational response induced by a government policy," aimed at increasing agricultural productivity and output through a sharp increase in the supply of soil nutrient. According to Kung and Cai, farmers welcomed this change, as chemical fertilizers rapidly showed more effective than traditional organic fertilizer in boosting crop yields.

Today, the scheme is still operating and keeps on having effects on the development of production capacities. Between 2002 and 2011, the national production of nitrogen fertilizers surged from 28 to 42 million tons in 2011 (FAO Database). The overcapacity of the sector provides farmers with products in abundant quantities at low prices, leading to over-consumption. The level of education of farmers and the lack of adequate soil diagnosis tools and expert teams on the field prevent Chinese farmers from balancing the volumes of nitrogen fertilizers (which represented more than 60 percent of the total fertilizer consumption in 2008 (USDA 2009)) with other types of fertilizers (phosphor and potash), leading to a low efficiency per kilogram spread.

The over-use of farm inputs poses several problems. Fertilizers spread in excess cannot be absorbed by the soil and leach into ground water, lakes, and rivers. The resulting pollution affects a growing urban population, which relies on these reserves for its water consumption (Jin et al. 2005).

In addition, the over-consumption of fertilizers has effects on agriculture itself, as over-consumption of nitrogen fertilizers can lead to an acidification of soils likely to lower their fertility without leading to higher yields. A study conducted by several Chinese and foreign institutes (Jua et al. 2009: 3045) estimates that "more efficient use of N fertilizer can allow current N application rates to be reduced by 30–60 percent."

Finally, the over-use of pesticides increases the amount of residues found in food. Chinese media regularly drive public attention toward this issue. At the end of the year 2011, Luo Xiwen (quoted by Liu and Zhang 2011), an academician from the Chinese Academy of Engineering, stated that "in some vegetables and fruits, up to 13 percent of pesticide residues are found, and heavy metals concentration exceeds quotas by 24 percent, and nitrate by 12 percent".

Concerns about environmental and food safety issues caused by unsustainable agricultural practices thus do exist. However, this does neither mean that these issues are brought to the political agenda nor that they are efficiently answered. Administrative and scientific circles as well as the civil society are constrained by a multiplicity of factors that prevent them from putting environment-friendly agriculture at the agenda or from taking effective action to address the environmental degradation caused by or affecting the agricultural sector. The capacity of Chinese administrative bodies to bring environmental issues to the agricultural agenda is constrained by two main factors: the fact that local governments are usually incited to limit the dissemination of information—and particularly of the information that could spread concern or alarm among the population—and their institutional fragmentation. These issues led, in the past, to important communication failures for major health scandals. A well-documented case that illustrates this point is the SRAS outbreak in 2003, which infected thousands and killed hundreds of people. According to Thornton (2009), the Chinese authorities could have anticipated, if not predicted, the appearance of a SARS-type epidemic. However, the author argues that the administrative fragmentation and the lack of coordination severely impaired an early and effective official response to the outbreak. In addition, Thornton denounces the responsibility of lower-level officials, who "intercepted and distorted the flow of information to upper levels, fearful that their perceived mishandling of the situation might result in negative performance evaluations." Finally, she argues that the emergence of a national crisis, in a way, contributed to the reinforcement of the power of the Chinese state, thereby implying that this latest would have a vested interest in crises. To our knowledge, such in-depth analyses were not conducted for food safety crises. However, similarities exist in the way disease outbreaks and food safety crises are handled, as both are related to health issues. In addition, numerous elements that can easily be found in the daily press suggest that the government might be reluctant to release information on food safety issues—the same way it was reluctant to release information on the SRAS outbreak—because of the damages that such information can have on the food sector and because of a certain willingness to maintain social stability.

Chinese consumers are perfectly aware of the fact that they lack information on the safety of food products. Hidden information, along with the fact that consumers are aware of this issue, led to the emergence of suspicion and sometimes to the rise of fiercer waves of panic. The lobby of industrial players adds up to the holding back of information by the government. A series of surveys conducted by the Pew Research Center in 2008, 2012, and 2013 (Kohut and Wike 2013) clearly illustrates the worsening of this suspicion, by evidencing that the topic of food safety moved to the forefront of people's concerns over the past few years. In particular, the series shows a tremendous increase in the percentage of people thinking that food safety is "a very big problem" in China between 2008 and 2012—as a consequence of the 2008 melamine scandal. Since then, the issue has remained among the top concerns of the population, just behind the issue of inflation, the corruption of officials, the gap between the rich and the poor, and the pollution of air and water (Fig. 3.7).

Food safety, in China, is not just about unsustainable farming practices relying too heavily on pesticides. Since the 2008 melamine milk crisis, the

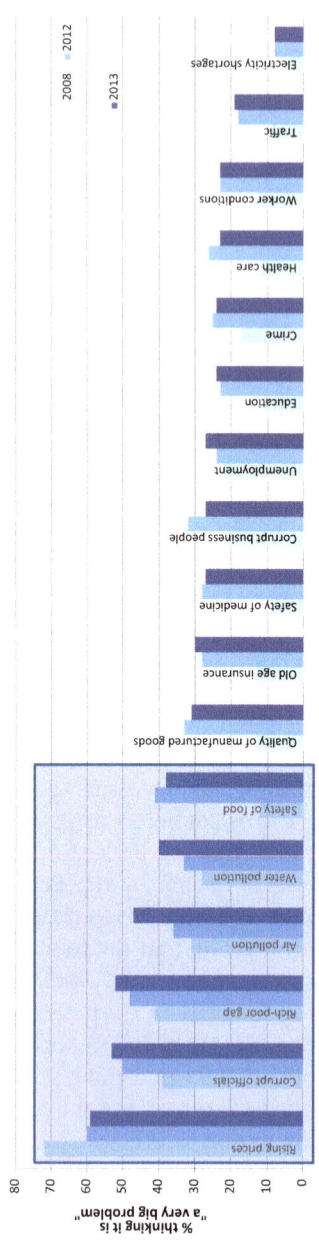

Fig. 3.7 Issues Chinese people consider as "very big problems" (Source: PEW Research Center Global Attitudes Project)

media regularly report food safety scandals linked to the race for increased profits. Examples are numerous, from clenbuterol meat to "recycled" cooking oil and rotten buns. Faced to the worsening of the situation, the government, over the past few years, was particularly active in reforming the system, issued a number of new rules and regulations and set up innovative information mechanisms. The Food Safety Law, for instance, was issued in 2009 in the aftermath of the melamine milk crisis. It strengthened regulations, controls and punishments, and created Internet and hotline early-warning mechanisms, enabling consumers to bring food safety issues to the attention of authorities. However, the political power and the scope of action of local bodies responsible for the compliance of enterprises remain low, partly because they lack human and financial resources[4] and partly because a portion of the revenue of inspectors comes from fines imposed on food producers (Duchatel 2011)—encouraging them to hide the information linked to the violation of regulation, as a way to safeguard their source of income.

In 2010, the Food Safety Commission was created. Among others, the Commission aims at addressing the overlaps in responsibilities of government bodies. However, in spite of successive reorganizations, overlaps persist. Several administrative entities are in charge of drafting policies linked to the agricultural sector. Such an institutional fragmentation is also observed at the lower levels of the administration, responsible for implementing policies. Local officials usually work inside competitive environments and power games, which does not encourage them to cooperate with each other. In some areas I went to, local bureaus of agriculture were conducting agricultural development projects, while in others, similar projects were run by poverty alleviation bureaus, without effective coordination or even communication between bureaus.

In addition to such institutional constraints, local governmental actors experience difficulties in implementing solutions for "greener agriculture." Faced to the degradation of its environment and to the consequences it had (and will have) on the national development (in terms of economy, health, food safety, social stability, etc.), the government has been particularly active in developing and promoting environmental protection policies over the past few years. However, in spite of a real willingness to improve the situation, the country still suffers from acute environmental issues. In fact, important obstacles arise when it comes to the implementation phase. These latest were explored by a large number of scholars. Most of them blame the rapid economic development and urbanization of China (Economy 2007), the "insufficient authority and the lack of co-ordination between institutional actors" (Jahiel 1998) and the administrative fragmentation and the defaults of the cadres evaluation system (Zhou et al. 2013). An important number of interviews conducted in the framework of this research showed that similar difficulties, linked to the administrative fragmentation and to the defaults of the cadres evaluation system, were impeding the implementation of environmental protection policies in the agricultural sector. This is consistent with the findings of Burns et al. (2010), for whom the inefficiency of the National Commission for Food Safety is due to the characteristics of the national cadres evalua-tion system, which pushed local officials to focus on issues for which they were assigned targets (e.g., social stability or economic growth), whereas other topics, such as food safety, are considered as less important and let aside.

Another explanation for the complexity of the implementation of envi-ronmental protection policies in the agricultural sector is linked to the challenge of maintaining a certain degree of food self-sufficiency. Environmental protection policies are seen as potential *threats* for the achievement of agricultural production targets. Even among central authorities, environmental protection is still considered as potentially harmful for national food security. As a Chinese research fellow working closely with the MOA explained to me, when I asked him if current debates on agricultural policies were involving environmental protection:

> There are many discussions currently ongoing about agricultural policies. Food security is one of China's main concerns. We also talk about environ-mental protection, indeed. About environmental protection and food safety. But if we observe what happened in other countries, in Korea, in Europe, everywhere, we see that developed countries first solved their food security problems then designed environmental protection agricultural policies such as payment for environmental services, etc. It is a question of productivity. (Interview, Beijing, June 2014)

What about civil society? Chinese NGOs concerned with environmen-tal issues have flourished over the past decades. In the middle of the 1990s, faced to the inefficiency of environmental protection policies—due to their systematic undermining compared to economic development poli-cies—the government progressively created a space for civil organizations willing to alleviate these issues. "Environment" was categorized as a (rela-tively) a-politic matter—compared to religious or ethnical questions—or at least nonconfrontational vis-à-vis the established power (Ho 2007: 189), which was careful enough to set up control mechanisms that would regulate the activities of these new organizations (Schwoob 2013a). Tolerated and even encouraged by the government—powerless or lacking interest for environmental issues—environmental NGOs developed rapidly. Nonexistent before 1994, there are today thousands of environmental NGOs registered throughout the whole country. Strongly linked to governmental institutions, environmental NGOs form a consultative and supportive network for the implementation of environmental policies.

However, over the past few years, the civil society started demonstrating an eagerness to bring environmental issues to the political agenda through protests, outside of the classical channels of regulated environmental organizations.[5] Chinese citizens increasingly take up such issues on the web. Progressively, the media, a number of active NGOs and the public opinion increasingly urged the government to better address environmental issues (Balme and Tang 2014). Rural issues also mobilize the interest of the civil society. The protests of farmers expropriated by local governments usually meet with the approval of other groups of the civil society and even with the approval of the central government. Land rights are today considered as *hefa quanli* (合法权利), or "legitimate rights". Civil society organizations engaged in the field of environmental protection thus do exist and are supported by a civil society increasingly aware of environmental issues. In addition, social actors know about rural issues and sometimes back farmers in their combat (in particular, against abusive land requisitions). However, NGOs bringing environmental issue *on the agricultural agenda* were almost nonexistent at the time this research was conducted. Environmental issues targeted by NGOs are usually the most visible forms of pollution such as air pollution or waste. In the agricultural sector, environmental protection remains a secondary objective and NGOs undertaking actions in the field of agriculture usually do so in order to address low economic development issues. In rural areas near Chongqing, the NGO I met was working on agricultural development essentially in order to address local poverty issues:

[For our agricultural development project] we selected the county of [...], which was close to Chongqing and to the market, and then we selected the poorest villages with the help of the county bureau of poverty alleviation. (Interview with the regional director of the NGO in Chongqing, October 2013)

According to a former NGO—now a micro-credit enterprise—working on agricultural development in Ningxia, the main part of microgrants is used by farmers to buy water, pesticides, and fertilizers—environmental issues alleviation was never mentioned during the interviews.

It happens that NGOs focus on environmental issues occurring in the agricultural sector. Greenpeace (2013), for instance, published a report revealing that herbs used for Chinese traditional medicine contained high concentrations of pesticides, likely to have harmful effects on human health. Although the report was widely cited in the media and profoundly shocked the public opinion, few reports are published on similar topics on a regular basis. In addition, actions undertaken by NGOs remain limited to research and reports and barely include field action.

The "connected" urban middle class, on its side, although backing farmers in theory and convinced that their demands are legitimate, are rather unlikely to take up their cause and to make efforts to bring this topic on the political agenda. For most of the social actors, rural areas and agriculture are rather far from their daily concerns, as illustrates this quote from someone in charge of raising funds for a foreign NGO conducting poverty alleviation projects in rural areas:

> It is very difficult to raise funds in China. It is very difficult because you don't have a status that authorizes you to raise funds publicly[6] [...] so you have to go from place to place to raise funds. And it's very difficult here also because people really don't care about the poor people in [the rural area where we are conducting projects]. They don't even know that there is a poverty line in China and they don't know how much it can be. Usually, in China, people mobilize during catastrophes. For NGOs dealing with seism or things like that, they get money. But not us, not really. (Interview, Beijing, November 2013)

However, concerns about the safety of food products kept on rising among consumers. These latest started developing individual strategies to curb the potential effects of unsafe food on their health. Some simply buy food stamped with organic or green labels. Others launch businesses in organic agriculture so that their children, family, and friends can eat safe food. Such strategies developed particularly in the wealthiest and more environment-conscious cities such as Beijing and Shanghai. When I asked a woman why she had decided to create an organic market in Beijing, she told me that when she gave birth to a little girl, she started having concerns about the safety of products she was feeding her with (Interview,

Beijing, April 2013). A manager who had just created an organic farm in the suburbs of Beijing told me a similar story:

> More than ten shareholders invested in this project. They gave a couple thousand yuan each. At the beginning, it was mostly friends, who wanted to grow their own fruits and vegetables, in order to ensure their food safety. Even if products we grew are not "organic," at least they are better than the ones we find on markets. (Interview, Beijing, April 2013)

Fieldwork showed that on order to cope with the issue of food safety, citizens usually preferred to establish individual strategies rather than taking part in collective action—an individualism that is very similar to what was observed for the daily tactics used by Chinese citizens to cope with environmental hazards (Tilt 2013). As a consequence, environmental issues caused by unsustainable farming practices did not lead to the development of NGOs taking effective action, as it was the case for industrial pollution.

Could research act to produce and disseminate more sustainable agricultural practices? For most of the Chinese leaders, productivity is still the main objective agricultural policies have to fulfill. In line with this objective, technological innovation is considered as a key lever to raise agricultural production, and important efforts—mainly made by the state (Zhang et al. 2011)—were dedicated to the development of research capacities over the past decade. Public investment devoted to agricultural research and development rose tremendously in the 2000s, and a significant amount of these expenditures was dedicated to research in biotechnology (Rozelle et al. 2005: 93) (Figs. 3.8, 3.9, and 3.10).

The research facilities I had the opportunity to visit were impressive. Laboratories were well equipped and using cutting edge research equipment.[7] Within the institutes I visited, I met a number of fellows carrying out research aimed at addressing environmental issues in the agricultural sector. A lot of researchers were working on genetics, but not all of them. Solutions also included drip irrigation, solar greenhouses, nonchemical pest control methods and systems, and nonchemical fertilizers (biochar, in particular). For a number of these solutions, China was considered as quite advanced on the topic on the scale of developed countries (Interviews, Beijing, November 2013, and Shanghai, March 2013). Moreover, a lot of researchers working in the institutes I visited were quite close to administrative bodies in charge of drafting policies. A number of

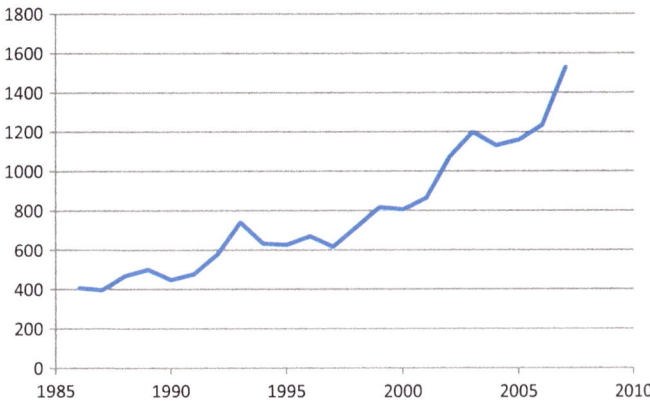

Fig. 3.8 Total agricultural R&D spending, public sector (million US dollar 2005) (Source: ASTI Database)

Fig. 3.9 Technicians sorting seeds in a gene bank in Beijing (Photography by the author, Oct. 2013)

Fig. 3.10 Researchers at work in a state key laboratory in Beijing (Gene sequencing for wheat, maize, rice, and soybean) (Photography by the author, Oct. 2013)

them even told me that they were asked by government officials to submit drafts for agricultural policies on certain topics. As insights from fieldwork seem to demonstrate, scientific circles had the means to bring the topic of environmental protection to the political agenda of agriculture.

However, as far as implementation is concerned, the task proves to be more difficult. Scientific circles are usually quite disconnected from farmers, with whom they have few opportunities to share their expertise. Exchanges though do exist between researchers and farmers. A wide network of extension services was established throughout China, and trainings provided by scientific staff are regularly proposed to farmers. Several bodies are involved in agricultural trainings. Apart from universities (e.g., the Chinese Agricultural University [CAU] in Beijing) and vocational schools (e.g., the Beijing Vocational College of Agriculture), the China Agricultural Broadcasting and Television School (CABTS) and the National Agricultural Technology Extension and Service Center (NATESC) are the two main national organisms in charge of agricultural training. The CABTS, which is under the direction of the MOA but is also supported by 21 ministries and commissions, offers graduate education for students as well as trainings for active farmers, thanks to a wide network of local schools.[8] Although the CABTS also provides rural areas with technological extension services, agricultural extension is rather the prerogative of the NATESC. The NATECS, also working under the direction of the MOA, supervises between 200,000 and 300,000 trainers across the country.

In spite of the wide network of local schools and extension service centers and the considerable number of trainers, people interviewed in rural areas expressed vehement criticisms of the system. Interviewees denounced the lack of knowledge of employees in extension service centers, their lack of interest in agricultural development and even their lack of direct contact with farmers. In a remote rural area in the municipality of Chongqing, I was told the local technical experts preferred to rely on hotlines. The local NGO I was visiting complained about the inefficiency of hotlines. According to the staff, farmers are reluctant to call people they do not know personally, even when they face important issues for which they know technical experts can give them advice:

> [Farmers' instructors of the township government] conduct trainings with the farmers, but the problem is that they usually don't follow up. But you have to follow up. If you simply conduct trainings, then the farmers won't follow the new methods. But the problem is that they don't have enough staff to follow everyone. For example, they set up a hotline for farmers so that they could call in case their pigs had a disease. But the thing is, they didn't call. But when we [NGO's employees] established our office in [the village], they came to see us for the diseases of their pigs. So we told the government to come sometimes, and now that they have come, farmers call them. (Interview, Chongqing, October 2013)

This quote illustrates the importance of personal connections in China (关系 *guanxi*), something already emphasized by a vast amount of research (for *guanxi* analysis in rural areas, see: Yan 1996; for the importance of guanxi among farmer-migrant communities, see Froissart 2007). In the vast corpus of literature exploring this topic, scholars emphasized the importance of maintaining social connections as a way to obtain financial and other resources. In the case of farming however, the significance of personal connections goes beyond accessing resources: even when farmers have the mean to be provided technical advice, they are usually reluctant to ask for it until they know personally the person providing advice. Several factors explain the situation, apart from the cultural significance of personal connections. Lack of insurance coverage and scarce personal financial resources increase the risk, for farmers, to bear the costs of following improper advice. It is easy to understand how trust plays a key role in the process. The distance usually put between farmers and advisors of agricultural extension services centers make the former skeptical about the good intentions of the latter, explaining the lack of efficiency of distant top-down training methods.

The distance put between trainers or technical advisors and farmers is also put between scientists and farmers. When I was visiting a "model farm" (or demonstration site) in the suburbs of Beijing, the guide told me:

> We are a window between China and the world. We want to show advanced agriculture science and technology to governments, enterprises. Even farmers come here to learn. (Interview, Beijing, October 2013)

The words "*even* farmers" reflect in fact a widespread situation among agricultural demonstration sites, where most of the displayed techniques are completely unaffordable for the vast majority of Chinese farmers. For instance, the price of a glass greenhouse such as the ones found in most of the demonstration sites I visited[9] was around 2000 RMB per square meter. To this had to be added the price of technologies such as hydroponics or vertical agricultural technology, also very popular in demonstration sites, given the stake of land scarcity. Finally, the price of water (120,000 RMB per year for a 4-hectare greenhouse) and electricity (1.2 million RMB per year for a 4-hectare greenhouse[10]) were also budget lines to have in mind. By comparison, in 2012, the average net revenue of rural households was less than 8000 RMB per year (National Bureau of Statistics).

3.1.4   Power Unbalance Between Food Enterprises and Farmers

Administrative and scientific circles as well as the civil society and NGOs are constrained by a multiplicity of factors preventing them from taking effective action to address environmental issues linked to agricultural production. Unbalanced power patterns between remaining players, food enterprises and farmers, explain why food enterprises have recently been favored by the government to address food safety and food security issues.

In the selected concrete system of action for this research—agricultural production at the county level, for selected agricultural subsectors—several broad categories of actors interact with each other. Usually, policy analysts consider three categories of stakeholders: public stakeholders, para-public stakeholders, and private stakeholders. These three categories, however, were insufficient to depict what I observed. Three other categories can be used by policy analysis: public actors; private actors belonging to the "hard core" of political space (e.g., interest groups or policy communities, as defined by Richardson and Jordan 1979; other useful references: Marsh and Rhodes 1992; Knoke 1996; Marin and Mayntz 1991; Wilson 2012); private actors of which the activities are more "subtle" in the political process (whether their silence is intentional or due to their lack of organization or resources). Three groups of stakeholders were first identified, which more or less match the earlier described categories: government officials, enterprises, and farmers. These broad categories comprise in reality a wide variety of players (Fig. 3.11). For instance, in food factories based in rural areas, individuals could belong to management teams (founders, CEOs, and vice-managers), be factory managers (in charge of supervising supply and industrial processes) or workers, and also contracted farmers and

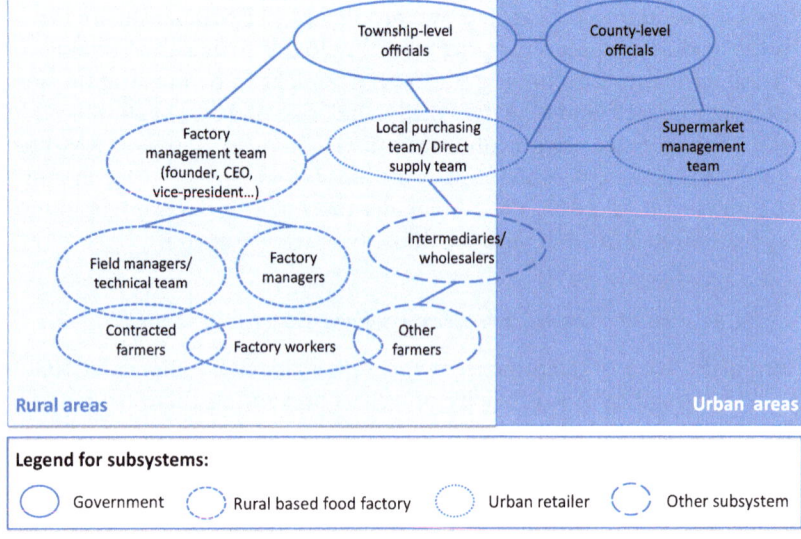

Fig. 3.11 Groups of actors of the concrete system of agricultural production

technicians working in fields managed to the factory. The additional complexity of the scheme comes from the fact that these groups of players overlap: for instance, contracted farmers are sometimes employed by local factories as workers, and, as such, belong to the group of factory workers as well. In order to simplify this scheme, three groups of players were delimited for the purpose of this analysis: local officials (from county and township levels), "enterprises" (by that, we mean mostly the factory management staff), and "farmers" (including farmer-workers). Individuals from each group of actors were interviewed in order to get a precise idea of their capacity to control the uncertainties of the system of action. In Crozier and Friedberg's methodology of organizational analysis (1977), the capacity of actors to control four different kinds of uncertainties grant them with increased power over the other actors of the system: (i) uncertainties linked to expertise; (ii) uncertainties linked to the environment(s) of the concrete system of action; (iii) uncertainties linked to communication and information; and (iv) uncertainties linked to the existence of organizational rules.

Expertise is one of the areas of uncertainty mentioned by Crozier and Friedberg. In the field of agricultural production, "expertise" is a combination of several kinds of knowledge, linked to agricultural production (the ability to grow crops or to raise livestock), and also to marketing, finance, and a variety of other related areas. The level of expertise of farmers is surprisingly low compared to rural-based enterprises. The vast majority of farmers indeed never had any vocational training and did not benefit from secondary or upper-secondary education. Agriculture, in China, is still widely a profession that does not stem from a choice made by individuals but is rather an inherited situation from which farmers usually try to escape. The intent here is not to say that farmers do not know how to grow products. According to most of the agribusinessmen I interviewed (who had usually never been farmers), *nongmin* have the expertise to farm, while they do not. For instance, the manager of a green farm in the suburbs of Beijing once told me:

> I majored in rural development in Renmin University. [But] Uncle L. [the farmer] has been working here for twenty years, he knows people, who make him good prices, he knows about manure. [...] We have several technical people on the farm. But in fact, Uncle L., as he has been growing vegetables for twenty years, knows better! We don't want to teach them. The only way to learn is through practice. It's the Chinese way: in the Chinese countryside, people are cooperating and learning through practice thanks to the advice of the elderly. (Interview, Beijing, April 2013)

Similarly, agricultural entrepreneurs, in Jiangxi and Shandong, were relying on farmers to grow products. They were barely seen in the fields and admitted that they were not capable of planting seeds, taking care of the land or harvesting themselves. However, considering the growing importance of environmental issues caused by intensive agricultural practices, agricultural expertise now includes environment-friendly practices and new technology. Chinese farmers are generally smallholders with little access to the newest technology that would help them modernize their farms. Enterprises, on the other side, have the financial capacity to hire technical staff with a certain level of expertise in agricultural technology. As a quote from an organic retailer in Beijing illustrates it:

> In order to deal with food safety, I only talk with businessmen, and not with peasants, because only businessmen have the money to do that and I wouldn't have time to manage every farmer. The small farmers don't have money, they just look at what is working and what is not working. (Interview, Beijing, November 2012)

Knowledge related to farming techniques is only one aspect of agricultural expertise. Other aspects include knowledge related to marketing and sales. In Crozier and Friedberg's methodology of organizational analysis, this kind of knowledge would not be labeled "expertise" but would rather refer to the control of "downstream environments" of agricultural production. Having access to the downstream environment of buyers is essential for producers. The range of buyers of agricultural products goes from food-processing companies to individual brokers, wholesale markets, retail markets, and individual consumers. However, the remoteness of farmers, their little connection to cities, and the fact that they often do not possess any vehicle make them lack control over this downstream environment. Local food-processing enterprises, on their side, have the possibility to recruit specialized marketing teams, who have the resources to look for buyers.

Another consequence of the isolation of farmers is that they have little control over information—the third type of power source depicted by Crozier and Friedberg. In spite of the active efforts of the government to develop "agricultural informatization", national and local information systems were still poorly developed for agriculture at the time I conducted fieldwork. On the opposite, enterprises, who are in touch with a wide population of sellers and buyers of agricultural products and inputs and have the technological means to access information, usually had a better knowledge of prices and on subsidy schemes.

In addition to expertise and to the downstream environment of agricultural production, another fundamental uncertainty of agricultural production, key to depict local patterns of power, is its upstream environment, and, in particular, access to credit. Financial resources are not only necessary for the modernization of the sector (through mechanization, the use of better inputs, etc.) but also essential to control the risks characterizing the agricultural production sector, as they provide coverage in case of bad harvest due to weather conditions, pests, or other events. Small farmers, however, considerably suffer from a lack of access to banking services. Since the middle of the 2000s, the need to reform the rural financial sector—partly to address this issue—was regularly emphasized both by experts and by government officials. In 2008, the central government issued specific demands regarding the reform: among other things, officials pointed at the necessity to modernize rural banking infrastructures, to improve credit availability, lessen credit conditions, and accelerate the building of mixed systems including commercial finance, cooperative finance, and governmental finance (Schwoob 2013b). Significant progress was made on the side of financial coverage. ATMs, retail points, and mobile phones payment services mushroomed throughout the countryside. However, in spite of these technological advances, credit availability and conditions are still tight for rural dwellers. According to an article from the Caijing magazine (Wang 2012), in 2009, only 32 percent of rural families had access to credit. According to interviews conducted between 2011 and 2014 for the purpose of this research, the situation did not evolve much since and access to credit is still limited for rural dwellers.

Farmers are particularly affected by the issue and are usually forced to use more informal channels to get loans (Zhou and Takeuchi 2010). The main reason behind this situation is the lack of eligible collateral, as farmers do not own their land and cannot use it to insure banks against liquidity shortfalls. According to Tang (2012), advisor of the State Council, "micro-credit" for small farmers is usually not considered as a profitable activity for rural banks, in comparison to loans to big agricultural enterprises. In addition, small farmers are scattered in the countryside and usually live in remote areas, which considerably increases the operating costs of rural banks. On the opposite, enterprises benefit from a much higher degree of trust in the banking sector. In addition, they can be backed by local governments. As a food-processing enterprise in Shandong told me:

> The government can appoint people to provide us various services, including financial services, discounted interest rates and access to preferential loans. (Interview, Shandong, November 2012)

To conclude, what the fieldwork of this research demonstrated is that strong uncertainties exist for agricultural production. These uncertainties are linked to the need for expertise, to the scarcity of information, or to the uncertainties in upstream and downstream environments. Food-processing enterprises based in rural areas control these uncertainties better than farmers (Table 3.1). The amount of time and human resources that a filling of gaps between enterprises and farmers would imply is enormous. In addition, the fact that farmers are scattered among the countryside and their remoteness considerably increases the transaction costs of trainings. The scale of the task is an important factor explaining why local governments usually prefer to rely on rural-based food-processing enterprises to drive the modernization of the agricultural sector, which is considered as an urgent task to alleviate food security and social stability issues.

Even in agricultural sectors where the involvement of the state has remained strong, such as grain production (through their involvement in state-owned farms,[11] mainly concentrated in the North-East, North-West, and subtropical areas (Zhang 2010), and through their involvement in the national storage system that mostly targets grain), the role of private stakeholders is keeping on rising. According to the National Bureau of Statistics, the grain output of state farms, in 2012, was 33.71 million tons, among a national grain output of almost 600 million tons. State farms thus only account for around 5 percent of the national grain production. The role that state farms have played in grain productivity achievements over the past few years was indeed limited to a small number of grain crops. On my fieldwork in Southern China, I could observe that rice was clearly out of this model of state farms leading the modernization of the grain production sector. In fact, what I observed for the rice production sector was quite similar to what I had observed in the fruits and vegetables sector. In Anhui and in Jiangsu, I met a lot of farmers still cultivating rice according to traditional methods, including traditional farming practices—usually labor intensive farming practices (some of them still relied on buffalo plough in Anhui province)—as well as informal exchanges of land, tools, labor, and small machinery (Figs. 3.12 and 3.13).

Table 3.1 Patterns of power related to the control of uncertainties of the concrete system of action of agricultural production

Stakeholder	Control of agricultural production		Control of downstream environment	Control of information		Control over production resources
	Expertise linked to agricultural activities	Expertise linked to agricultural new technologies or sustainable practices	Access to markets	Information linked to prices of agricultural products	Information linked to prices of inputs	Financial resources
Enterprises	−	+	+	+	+	+
Farmers	+	−	−	−	−	−

Fig. 3.12 Farmers going to the fields with buffalo, Anhui (Photography by the author, June 2014)

Fig. 3.13 Farmers transplanting rice by hand, Anhui (Photography by the author, June 2014)

Investors were rare in this picture but not absent. I had the opportunity to meet some of them, especially in Anhui and Jiangsu provinces. According to the interviews I could conduct, their development models are very similar to what was described earlier in the text: investors were usually coming from outside of the farming sector, after having managed to gather funds and to establish relationships with local governments to obtain land and hire local farmers to grow rice (Figs. 3.14, 3.15, 3.16, and 3.17).

3.2   UPSTREAM AND DOWNSTREAM ACTORS INVITING THEMSELVES IN THE PICTURE

Food-processing enterprises based in rural areas are not the sole stakeholders brought to the forefront of agricultural modernization. Retailers established in urban areas are also increasingly encouraged to engage in concrete actions to take part and were present on the fields. Through the upstream integration of retailers, the government hopes to address inflation and food safety issues without penalizing consumers. Intermediaries of the food chain, on the opposite, make easy scapegoats, as "people know

Fig. 3.14 Farmer-worker at work for a rice growing company in Jiangsu (Photography by the author, June 2013)

that they are everywhere but nobody really knows them," as formulated by a manager of a food-processing enterprise (Interview, Jiangxi, October 2012). In addition to the integration of downstream actors up in the food chain, upstream actors—such as agrochemical companies—sometimes take part in agricultural modernization as well.

3.2.1   Bringing Retailers Up in the Chain Through Direct Purchase

In 2007, the Ministry of Commerce gathered the CEOs of the nine biggest retailing companies in China for a special meeting, which marked the official launch of the model of "farmer-supermarket direct purchase" (DP). Under this model, retailers are encouraged to purchase agricultural

Fig. 3.15 Women farmers-workers working for a rice growing company taking a break in Anhui province (Photography by the author, June 2014)

Fig. 3.16 Local governments' officials and rice growing company's managers, Jiangsu province (Photography by the author, June 2013)

Fig. 3.17 Mechanized rice-transplanting (Rice growing company, Jiangsu province) (Photography by the author, June 2013)

products directly from producers in rural areas and are actively supported by the Ministry of Commerce in their "DP" initiatives through tax abatements and other incentives. In 2011, 2000 supermarkets had already developed DP projects (Hu 2013). The model is expected to help the government reach its objectives in terms of rural development and inflation. DP is indeed supposed both to increase and stabilize farmers' income over time—by linking them directly to final markets and enabling them to sell their products at relatively stable prices—and to curb food-price inflation—by deducing the margin of intermediaries from final food prices. In addition, DP projects are expected to be attractive to retailers (as DP aims at reducing purchase costs and at improving the safety of products) (Hu 2013) and to benefit final consumers as well in terms of food safety.

DP projects in fact already existed in China prior to the official launch of 2007. Foreign food distribution enterprises, in particular, were familiar with these methods, as they were already using them in other countries. In order to understand the reasons of the development of DP in China before 2007, it is important to look at the business environment of retail enterprises of the past decade. In the 2000s, the Chinese retail sector underwent a phase of accelerated development. Supermarkets and hypermarkets rapidly burgeoned throughout the whole country, starting with the developed areas of eastern China. Supermarkets emerged as important players in an environment traditionally dominated by small retailers, grocery stores and marketplaces. Today, the retail sector has become highly competitive. Profit margins are small, pushing retailers to engage in mergers and acquisitions. These past few years saw a real concentration of the sector, which consolidated a limited number of players. In the developed provinces located in the East of the country, the multiplication of players and the growth of several big retail companies dramatically raised competition. At the end of 2011, when I began interviewing stakeholders in the retail sector, the saturation of city centers and the rise in real estate prices had already started pushing supermarkets to suburban areas or to the western parts of the country, where they were much more warmly welcomed by local governments—even though today, competition is already emerging in these areas as well, almost as fiercely as in the city centers of eastern developed China.

In such a competitive environment, the price of products sold by supermarkets comes under strong downward pressure. Retailers face the local competition of a multiplicity of small players who can rapidly take measures to outweigh the hard discount strategies implemented by retailers (Interviews with supermarket managers, Shanghai, May 2012). In addition, profits can be increased neither by lowering rental costs—as supermarkets have to compete with the other users of urban land, which keeps on pushing land prices up—nor by decreasing salaries—as low wages are already the rule in the sector. In the past, dealing with "brokers" enabled supermarkets to negotiate low prices for food products, as brokers selling a wide variety of products could "invest" in food while making a margin on other products with higher value-added. As was telling me a former sales manager of a supermarket in Shanghai:

> The main interest of brokers is that they sell apples, but also nuts, vacuum cleaners [...] As a consequence, a broker can 'invest' in apples by selling them to us at 1 yuan instead of 2 yuan, and he knows that he will be able to make profits on other products. (Interview, Shanghai, June 2012)

Given the large quantities of products they trade, supermarkets and brokers can easily achieve economies of scale. However, faced to the increasing competition in the retail sector, supermarket companies had to find another way to maintain their margins, either by trying to reduce costs or by differentiating themselves from their competitors to attract more customers.

DP appeared as a way for supermarkets to implement both of these two strategies at the same time. Food-processing plants established in rural areas, thanks to the recent support of local governments, are modernizing and are becoming increasingly able to answer the specific demands of supermarkets. In addition, as food safety is rising among the concerns of customers, building brand images based on the quality and safety of food products is viewed as an interesting way to improve one's position in the competitive environment of retailers. As was saying a manager working at Y., an important international retailer in Shanghai:

> Among the basic requirements, there is of course the fact that it has to be clean and tidy. But there is also food safety. In fact, we have to define a marketing position: 'Tomorrow, I want to be recognized by my clients for food safety.' (Interview, Shanghai, October 2012)

At another international retailer, X., with Chinese headquarters also established in Shanghai, where I conducted several rounds of interviews from October 2012 to October 2013, I was told two kinds of DP projects were running prior to 2007: (i) DP projects focusing on producers in rural areas as a way to reduce food prices; (ii) DP projects selecting producers in rural areas in order to improve the quality of products. In 2013, X. decided to restart a former DP project, aimed at improving the quality of products. According to the people I interviewed, several reasons motivated this choice. The official reason was that X. was willing to meet the ever-increasing expectations of its consumers in terms of food safety and in terms of freshness and taste of agricultural products such as fruits and vegetables. The other reasons were linked to the recent impetus given by the government: namely, all the actions that were conducted by central and local government officials since the official launch of the DP model in 2007. The following quote illustrates this point:

> The Mofcom recently went with X. to visit quality lines in Brazil. After this journey, the Chinese government asked X. to restart its quality line project, and specifically in Jiangxi province, which is the birthplace of the Chinese Communist Party and for which it is necessary to develop economy and create a good image. (Interview with the new manager of quality DP projects, Shanghai, October 2013)

The willingness of a number of government officials to push X. to launch DP projects in fact dates from before this visit. According to a manager of DP projects at X., the initiative was taken by the government in 2007, in order to contain inflation:

> They asked us to make an effort on our profit margin. However, on staple products like spinach, potatoes or tomatoes, our margin is close to zero, and it was impossible to "make efforts". But it was true that from the farm to the supermarket, there usually were 5-6-7 brokers, who were taking margins. We thus started implementing direct purchase projects, in order to have a better traceability, a better food safety image for the consumer, and also to have products at better prices. (Interview, Shanghai, October 2012)

DP projects involve to completely rethink traditional food supply models and to gradually replace brokers with rural suppliers, product by product. This task is quite challenging and time-consuming for retailers. In order to encourage them to engage in DP projects, the central government set up tax abatements (there is no VAT for products directly purchased from farms) as well as special business licenses for a number of rural producers, which facilitate the process. In addition, government officials are sometimes deeply involved in such projects. For local officials working in rural areas, being involved in DP projects can be a way to promote enterprises with which they have close links—either for personal, political, or economic reasons. As was explaining a manager of DP projects at X.:

> It goes that way: the government tells us: you will work with this supplier, with this one here, with this one there. You will work with this slaughterhouse. It is all informal obligations of course. However, if we do not do it, we face the risk to find something [bad about us] in the media the week after. The slaughterhouses are linked to the government. There is this guy who managed to get the right contacts, or sometimes slaughterhouses belong to officials. It works that way. [...] When we started to work with farmers, it was mainly with very integrated farms, which controlled the whole chain from production to transformation, as it was easier for us. They were of course very close to the government. (Interview, Shanghai, October 2012)

It is also a way, especially for higher-level officials, to build an image of a politician concerned about his country and going to the fields, both to look good toward above-level officials who will eventually evaluate them and to increase their legitimacy among consumers-citizens—something that has become particularly important given the rise of the stakes of food safety. As was saying the new manager in charge of quality DP projects at

X., expressing her embarrassment vis-à-vis the presence of high-level officials on the field:

> Our first sourcing in Jiangxi for the quality lines was extremely political, as the government [Mofcom] was with us. […] When I was in Fujian, I was escorted by the local contact from the MOA, by the deputy governor of the township and the director of economic development. […] For our second audit, the government again expressed the wish to come with us. (Interview, Shanghai, October 2013)

Is the greater involvement of retailers likely to compete with the actions undertaken by food-processing enterprises in rural areas? This is what would theoretically happen if retailers were launching DP projects with farmers. However, insights from fieldwork demonstrated that this was usually not the case. At the beginning, X. wished to work directly with farmers, as this is what the company usually does in other countries. As farmers were too small and could not produce volumes likely to meet demand, X. started looking for farm associations. However, the company's attempts rapidly proved that the experience was going to be difficult to conduct in China, and X. rapidly turned to food-processing enterprises based in rural areas. The main problem of farm associations mentioned by X. was that they were not able to closely control and monitor farming practices:

> [For the implementation of the new quality line,] we wanted to go back to the previous system of farm associations. The idea of the system was that farmers would join forces and exchange good practices and knowledge, and that we would manage sales activities, it was about to bring huge benefits. However, the thing is that big farmers had put small farmers under their control and that they were sometimes 50–100 farmers belonging to the same farm association. As a consequence, there was no monitoring of agricultural practices. (Interview, Jiangxi, October 2013)

As X., other retailers are usually more eager to turn to food-processing enterprises established in rural areas for DP projects. This does not mean that farming practices are systematically controlled by food-processing enterprises. In China, traceability usually stops at the factory level:

> We have been working on traceability with the government for over two years. […] Actually, it is not possible to implement a complete traceability

system, not today, not given the current production conditions. Traceability systems can be set up only from slaughterhouses to supermarkets. (Interview with manager of DP projects, Shanghai, October 2012)

When I went to Jiangxi to conduct interviews in orange factories, I could see the challenges factories had to face in terms of traceability from the farmer to the gate of the factory (Fig. 3.18). In fact, food-processing enterprises usually know which supplier oranges come from when trucks arrive at the factory gates, for the simple reason that they have to know who they have to pay (Fig. 3.19). However, suppliers can be farmers with identified fields and farming practices but can also be farm associations (gathering a wide number of farmers with different farming practices) or even brokers picking up oranges in several farms or buying them from other brokers before coming to the factory. Requiring factory suppliers to know which farm their oranges come from would take time and jeopardize the ability of the factory to answer the growing demand of its clients and, as a consequence, threaten its development.

In addition, implementing traceability from the field to the factory also implies changing processes inside the factory. Except from a few "high quality" products sold at a very expensive price, oranges coming from dif-

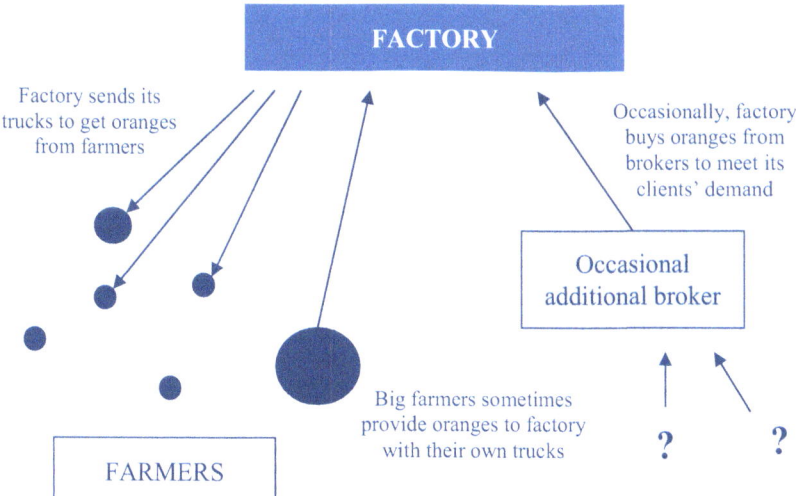

Fig. 3.18   Factory suppliers

Fig. 3.19 Workers unloading a broker's truck in Jiangxi (Photography by the author, Oct. 2013)

ferent suppliers are processed together (Fig. 3.20). Inside the factory, traceability is not just about putting stickers with barcodes on batches. It implies rethinking the whole processing system, in order to make sure that the right batches go to the right clients, that batches are processed and stocked without being mixed, and other things as well that require superior supply chain management skills and tools. The fact that small volumes are traded by each farmer and arrive to the factory gate further increases the complexity of the task.

Fig. 3.20 Orange-processing factory in Jiangxi (Photography by the author, Oct. 2013)

However, food-processing enterprises, compared to farm associations, usually *had the mean to convince* retailers that it was possible to implement traceability. Most of the enterprises I met indeed had their own plot, and it was easy to say that they would "reserve" a portion of their plot (where they could control farming practices) to supply the retailer's demand in terms of farming practices.

In reality, in a factory I visited, I saw that oranges packed for another retailer, Z. (which I knew had similar demands in terms of traceability and had agreed on these things with the processing enterprise) were coming from other plots not belonging to the enterprise[12] (and, as a consequence, where farming practices were not closely controlled). However, X. was still willing to take the risk, as managers believed in their capacity to convince producers that is was better for them to implement traceability systems, and willing to help them through trainings.

To sum up, DP projects, far from establishing direct links between retailers and farmers, only give greater power to already empowered rural-based food-processing enterprises. Retailers deal preferentially with factory managers, whom they try to train to traceability, while factory managers, on their side, try to implement monitoring systems to improve farming practices—at least in their own plots.

3.2.2   *Bringing Agrochemical Companies Downstream*

Apart from retailers, another group of players appeared in the fields over the past few years: agrochemical companies. It is not unusual for contemporary agrochemical companies to move downstream in the food chain. The moving of agrochemical companies downstream is not a new development either. At the end of the 1990s, the large transnational agrochemical companies, as a way to overcome the effects of a declining market for pesticides (Conway 2000), started investing in the development of transgenic crops. Not only did they invest in research and development, but they also started purchasing existing seed companies, first in industrialized countries and then in the developing world (FAO 2004), giving birth to "agrobiotechnology" companies, such as BASF, Syngenta or Monsanto. More recently, agrochemical (or agrobiotechnology) companies have increasingly been seeking to invest further down in the food chain, and particularly in the food-processing industry. As soon as the end of the 1990s, Heffernan (1999) had identified several clusters of firms integrated in various levels of the food chain through joint ventures. However, his analysis did not identify, at that time, conglomerates integrated all along the chain, whereas today, an increasing number of biotechnology companies are integrated "from gene to supermarkets shelves." The case of Limagrain well illustrates this process. Limagrain, at first a French agricultural cooperative gathering grain farmers ("Coopérative de Production et Vente de Semences Sélectionnées du Massif Central"), now ranks among the leading multinational seed companies, and is indeed running activities all along the food chain through its various subsidiaries, such as Jacquet, acquired in 1995, and Brossard, acquired in 2011, producing processed food and bread.

The biotechnology sector, in China, is though heavily regulated. Only a few genetically modified (GM) varieties of food crops are approved for commercial cultivation, such as a few varieties of tomatoes and peppers (since 1998) and of papaya (since 2006) (even though 80 % of the cotton cultivated in China is GM). Two matters of concern to the government prevent the development of commercial cultivation of GM food crops. The first is linked to the potential social protests that could arise if genetically modified organisms (GMOs) were placed on the market. Although a number of studies suggest that the majority of the population is in favor of GMOs (Zhang et al. 2010), debates are fierce on whether or not GMOs should be put on market shelves and in consumers' plates. Debates occur on online social networks among concerned consumers, and also among members of the government themselves. In August 2013, a major-general of the PLA and Deputy Secretary-General of China's National Security Forum, Peng Guangqian, published a tribune denouncing that GMOs were part of a military strategy perpetuated by the United States against China, in the following words: "Since the establishment of the PRC, it has already been proved that enemies could not use military force to conquer us. However, with this kind of subtle bacteriological weapon in the cards, we could lose our vigilance"[13] (Peng 2013). Following the publication of Peng Guangqian's article, the News Office of the MOA published an interview of an expert from the GMO security committee, answering the concerns expressed by the major-general and trying to reassure the population (Ministry of Agriculture 2013).

The second matter of concern is linked to the fact that a liberalization of the market of GM varieties would currently benefit mostly foreign companies such as Monsanto or Syngenta. Indeed, despite the tremendously high levels of public investment in the development of research in biotechnology over the past few years, foreign companies are currently still more competitive than Chinese companies for GMOs for a number of products such as maize. In order to limit the competitiveness of foreign biotechnology enterprises in GMOs, regulations limit the activities conducted by foreign biotechnology companies. These latest can conduct research activities on the Chinese territory but only through joint ventures with Chinese partners. The State Council indeed stipulates that foreign investment in the conventional seed industry is a "restricted" activity (foreign companies can only own up to 49 percent of a joint venture with a Chinese partner) and foreign company development, production, or marketing of transgenic plants in China is labeled as a "prohibited" activity (USDA 2012).

The market of pesticides and fertilizers is less regulated than the seed industry. However, it is highly competitive. In other countries around the world, the agrochemical market is usually concentrated among a limited number of multinational companies. For instance, six of the world's largest agrochemical and seed corporations (BASF, Monsanto, Bayer, Syngenta, DuPont, and Dow) control 75 percent of the global agrochemical market (ETC Group 2008: 14). In China however, the situation is radically different. The agrochemical market is heavily fragmented, with the "big five" (Nopoison, Syngenta, Bayer, Dow and Dupont) having collectively only 20–35 percent market share, the remaining 80 percent being owned by more than 2000 agrochemical enterprises, mostly local.

In such an environment, it is particularly difficult for foreign companies selling agrochemicals and biotechnology to acquire new market shares. Among the strategies deployed by international agrochemical companies, the one developed by "A." is particularly interesting, as it involves its integration downstream in the food chain. A. is an international company—among the biggest agrochemical companies worldwide—which established a Limited Company in China in 2000. The company now employs around 1000 employees in its agricultural business unit (the group, in total, employs around 13,000 people in China) and has one local production site. The company has been particularly active in developing strategies to reach new clients—farmers—through three main channels. The first channel is made of a network of sales representatives or consultants working for A. in local "agribiosolutions" shops. This network forms the basis of a rather classical marketing strategy. The second channel is more unusual, as it is made of four research and development centers based in rural areas. The goals of these "agrisolution centers" are to train farmers—potential new clients—and to promote A.'s technology through field demonstration. Agrisolution centers are established in partnership with local research centers, which provide resources to the company, such as land (for field demonstration), facilities, and sometimes technicians. The rationale of this strategy is twofold. The first rationale is market development. Demonstrations and trainings can help the company reach new clients or people able to put them in touch with potential new clients, such as opinion leaders and local officials. In addition, demonstrations and trainings can help the company develop a new business activity: consultancy. Instead of trying to sell more pesticides or fertilizers—hardly feasible considering the current consumption levels in China—developing consultancy represents an interesting and profitable alternative strategy. As was saying a manager in charge of "new business development" at A.:

> We do not want to sell products only. Our model is that we want to sell integrated solutions. (Interview, Beijing, November 2014)

The second rationale is linked to the attention paid by the company to establish and maintain good relationships with central and local government officials—one reason that was regularly mentioned by foreign retailers as well. Training farmers and contributing to the improvement of their living conditions contribute a lot to the company's recognition by the government. As was explaining the same manager:

> In order to sell them our products, we have to convince them that it is good both for food security and for food safety and that it will increase the revenue of farmers. This is a slide that we show to the Ministry of Agriculture.

It demonstrates that with our products, we can achieve increase yield by 13 percent, revenue by 14 percent, and decrease the use of pesticides by 62 percent. (Three concerns that strongly echo the goals promoted by agricultural policies.) (Interview, Beijing, November 2014)

Finally, the company is also increasingly developing partnerships with a wide range of "nontraditional" players, even more downstream in the food chain than farmers, such as retailers or food-processing enterprises (the third channel). The aim of this strategy is to convince these enterprises to buy food products that have been grown in the fields using A.'s agrochemical products or to encourage them to convince their suppliers to use A.'s products. A.'s argument is based on better food safety and the capacity to export (as there are strict regulations on residues for exports). For A., the rationale of this third channel is to reach more clients, either directly—for food-processing enterprises having their own farm bases—or indirectly. As explained by the manager in charge of new business development at A.:

> We try to link the actors of the food chain with our customers. In some cases these players have direct contracts with farms, and so we try to sell them our products. But it really depends on the kind of product we are talking about. For instance, we work with Mac Kain, which has major potato fields in Xinjiang. The size of farms is really huge, so even if a small number of farms buy our products, it can have a huge impact. [...] Because food companies and supermarkets have to ensure food safety, and in particular retailers, because retailers are facing the first impacts from the consumer side (in case of a problem of food safety), not the farmers, the farmers are very far. So we are demonstrating things to farmers *and* to food companies (in terms of stewardship, how they can protect their employees). We see ourselves as multipliers. (Interview, Beijing, November 2014)

Partnerships currently developed by A. even involve banks. The purpose of this strategic cooperation is to overcome farmers' obstacles in terms of access to credit. In the framework of such partnerships, A. offers crop solutions and figures demonstrating their potential impacts on farmers' income. The "scientific evidence" of probable future rise in farmers' income provided by A. reassures banks, unlocking credit access for farmers.

As A., agrochemical companies, in China, are likely to address the difficulties they experience in a particularly competitive environment by implementing a variety of innovative strategies, some of which include establishing partnerships with downstream players in the food chain. However, it is worth mentioning that A.'s model is quite recent. The time and cost needed to establish such strategies currently limits this possibility to a small number of players on the market.

3.3   CONCLUSION

Fieldwork conducted in Lushan and Lanshui, along with additional insight I could collect from interviews conducted in other areas (e.g., Jiangsu, Chongqing, and Beijing) and from actors running projects elsewhere in the Chinese countryside, showed strong evidence that local governments usually preferred to rely on rural-based food-processing enterprises to conduct agricultural modernization rather than on farmers (as it was the case for a number of countries such as France) or NGOs (as it was the case for environmental issues alleviation in China). This choice can be explained by several reasons. The first reason is rooted in the history of China. For the past four decades, local officials have kept on relying on enterprises to achieve local development goals, especially in rural areas. Officials are used to navigate in state-enterprises nexus and much more easily communicate with entrepreneurs than with farmers. Second, rural food enterprises are empowered by the current melting of food chains, artificially orchestrated by the state to limit the rise in food prices without negatively impacting (already low) farmers' income. Considering the fact that most farmers are still smallholders scattered in rural areas and sometimes living in remote places, cutting all intermediaries out of the food chain is hardly conceivable. As a consequence, food-processing enterprises established in rural areas remain nonremovable intermediaries between government officials and peasants. The third reason why local governments are eager to rely on food enterprises is linked to the urgency of the rising stakes at hand in terms of food security and food safety and to the attractiveness of the solution to rely on food enterprises. These latest indeed hold much greater control over a number of uncertainties characterizing the agricultural sector, such as expertise and upstream and downstream environments. The power and capacity of enterprises to steer agricultural modernization and to drive agricultural production toward more productive and more sustainable practices are fundamental factors that explain the rationale for the choice of local governments to rely on rural food enterprises to conduct agricultural modernization.

The role played by food-processing enterprises is increasingly supplemented by retailers, who were recently invited to take part in the process through the launch of DP projects. Networks of enterprises of the food chain consolidated across China, tying together big retailers with local food-processing industries. In addition, this role has also recently been supplemented, to a lesser extent, by upstream agrochemical companies. Both upstream and downstream players have an interest in expanding their area of activities in the food chain. Retailers, by establishing direct links with rural producers, lower the cost of agricultural products and build differentiation strategies focused on traceability, quality, and safety. Agrochemical companies, on their side, are likely to reach new clients by establishing direct contact with farmers, through the development of consultancy activities or good relationships with government officials, or indirect contact with farmers through links with "nontraditional" downstream partners such as retailers or food-processing enterprises.

The increasing reliance of local governments on food-processing enterprises and on other private players downstream or upstream in the food chain opened a new area of opportunities for these actors, leading to the emergence of firm agriculture and industrial and market-based private agriculture. At first rooted in rural areas, this type of agribusiness entrepreneurship progressively became more "transversal," straddling between urban and rural areas. Rural-based food-processing enterprises remained key and irremovable intermediaries in the food chain able to link the two worlds, as the sole players able to cope with the transaction costs of scattered suppliers and answer the specific demand of retailers.

The ever-greater involvement, in agricultural production, of private actors traditionally considered as downstream players in the food chain (e.g., food-processing enterprises and retailers) or as upstream private actors (e.g., agrochemical companies) resembles what was described by some scholars as the "privatization" of agricultural policies. Fouilleux (2010: 390), in her work on voluntary standards, demonstrated that private players, worldwide, increasingly have "the capacity to autonomously enact sets of rules" and that these sets of rules were "intended to apply to an important number of producers—if not all—and sometimes [became] reference points for public action, and [could] thus be considered as forms of *public private* policies".[14] She argues that people willing to fully understand the current regulations of the food sector have to go beyond the "classical" actors of agricultural policies (the state, the unions, and professional organisms), and take into account other players as well, such as multinational enterprises (MNEs), NGOs, banks, big retailers, and certifying bodies. These actors, who used to be out of the fields, now increasingly interfere in the drafting, implementation, evaluation and control of public food and agricultural policies (Fouilleux 2010: 389).

Although Eve Fouilleux bases her argument on the understanding of the formulation of food standards, other examples illustrate the rising role of food and agricultural firms. In the same book, "Les mondes agricoles en Politique", Goulet (2010) explains, for instance, how informal networks and professional associations, supported by agrochemical firms, started conducting experimental research on soil at the beginning of the 2000s. Their results invalidated the ones of experiments almost exclusively conducted by national scientific circles until the end of the 1990s, calling into question a whole set of farming practices that had been standardized with agricultural modernization.

According to Purseigle (2012), two fundamental changes marked agricultural modernization: the turning of peasants into farmers (from "paysans" to "agriculteurs") and the more recent emergence of "firm agriculture", or "corporate-style" farming, which includes, for instance, "corporate" farms, "capitalist-driven" family farms, or agricultural service supply agencies. For Purseigle, this last evolution—the emergence of firm agricultures—completely changed the place of farmers both in the Western world and in emerging and developing countries. Farmers associations and agricultural firms started lobbying and their lobbying capacity rapidly expanded well beyond the national scale.

Did the downstream and upstream integration of retailers, food-processing and agrochemical companies lead to the establishment of what Hervieu and Purseigle call "firm agriculture" in China? According to the authors, firm agriculture has two characteristics. Firstly, it is based on a multiplicity of decision-making entities, each with its own interests. Secondly, firm agriculture is widely relying on new, nonagricultural, tangible, and intangible resources (Nguyen and Purseigle 2012). As such, we can say that firm agriculture has indeed emerged in China. If we take a closer look at how retailers integrated upstream, we see that the government played a nonnegligible role in the process, through the establishment of subsidies and the personal involvement of a number of officials in DP projects and actions. However, enterprises have an interest in upstream integration outside the frames set by the government. Firstly, DP is likely to lower the cost of agricultural products, through the elimination of a number of intermediaries between farmers and supermarkets. Secondly, DP might help retailers build differentiation strategies focused on traceability, quality, and safety. Driven by their interests, which were complemented by governmental support, enterprises emerged all along the food chain, leading to the development of a privatized and market-based industrialized agriculture.

Rooted in rural areas, the earliest forms of this type of industrial market-based agriculture developed at the beginning of the 2000s. The development of private enterprises in the agricultural sector is very different from what was observed in the other sectors of the economy. In these latest, private entrepreneurship developed progressively, from SOEs to collective and private enterprises (Nee and Opper 2012). In agriculture (at least in activities such as fruits and vegetables production), the state brutally withdrew with the abolition of People's Communes and the implementation of the HRS, letting small farmers develop private entrepreneurship. At the beginning of the 2000s, local officials, already quite used to deal with private enterprises in the industrial sector, decided to encourage the development of "transversal" networks (in the sense of local state-enterprises networks) of food-processing enterprises in rural areas, in order to speed up agricultural modernization.

Compared to other countries such as France, where agricultural modernization was mostly carried out by farmers and agricultural cooperatives (and, in this sense, looks like the "capitalism from below" described by Nee and Opper (2012)), China rather took the path of a form of agribusiness entrepreneurship acting *on* farmers. Agricultural industrial private entrepreneurs progressively became liaison agents, at the same time levers for modernization and nonremovable intermediaries between government officials and peasants, who still have little connection with each other (at least, at the county and township levels).

Faced to the emergence of agri-industrial entrepreneurs, does the state control these new actors or are we witnessing a "privatization" of agricultural policies? Is the Chinese state threatened to become "delegitimized", as phrased by Peters and Pierre (1998: 225), "because of the control of information and implementation structures by private actors"? Is the increased participation of private actors a sign of a "retreat" or a "hollowing out" (Peters 1993; Rhodes 1994) or even a "collapse" (Botha 1999) of the state? Or is the state keeping on *administrating* public authority even if it is no longer vested with its monopoly? How is the state capacity evolving? Is it "declining in some areas and rising in others" (Ikenberry 2003: 351)? Or is the state able to create new frames for political action (Hibou 1998)? What is the Chinese state? The following chapter wishes to provide some answers to these questions.

Notes

1. "In an ideal-type corporatist system, at the national level the state recognizes one and only one organization (say, a national labor union, a business association, a farmers' association) as the sole representative of the sectoral interests of the individuals, enterprises or institutions that comprise that organization's assigned constituency. The state determines which organizations will be recognized as legitimate, and forms and unequal partnership of sorts with such organizations."

2. In 2012, the average size of land owned by rural households engaged in agriculture (农村居民家庭经营耕地面积) was 2.34 mu (0.15 ha). Other data estimate that the average farm size should be closer to 0.5 ha.

3. Harvesting is made before fruits are ripened, when they are less fragile. Fruits are then stored and artificially ripened depending on demand.

4. There were only 3900 laboratories to test the safety of food products in 2007, or one laboratory for more than 300,000 residents (The US-China Business Council 2007). An interview conducted in Beijing in November 2011 with an agent of the FAO working with food safety controllers confirmed that although progress had been made since 2007, local bodies were still lacking financial and human resources to efficiently control food safety.

5. In the summer of 2011, no less than three main demonstration episodes caused by environmental concerns occurred: in Dalian (Liaoning), residents demonstrated against the building of a chemical plant; in Haining (Zhejiang), city dwellers obtained the (temporary) closure of a solar panel factory; in Haimen (Jiangsu), a thermal power plant project had to be stopped because of protests.

6. As it is the case for most of the foreign NGOs operating in China.

7. An Illumina's HiSeq 2000 (sequencing equipment), worth between €500,000 and €1 million, was found in one state key laboratory in Beijing doing research in crop sciences (mainly on wheat, rice, maize, and soybeans)—I was told that as a National Key Facility since 2003, the laboratory received 200 million RMB per year to fund the contracted staff and some equipment (the permanent staff were paid as state employees) and could also apply for other fundings for equipment.

8. A total of 39 at the provincial level; 372 at the municipal level; 2071 at the county level; 10,805 at the village level (Source: CABTS presentation held during an EU-China meeting in Tianjin, in November 2012).

9. The name "demonstration site" is quite ambiguous. It can indeed be a technological park oriented toward enterprises (either to attract investment or to sell technology: 农业科技园 nongye keji yuan, "science and technology park"), an experimental base attached to a research center

(试验站 shiyan zhan, "experimental station"), a site promoting technology among a wider public (enterprises, entrepreneurs, teachers, political leaders, farmers), usually linked to the local agricultural extension service bureau (农业技术推广站 nongye jishu tui guanzhan, "agricultural technology promotion station"), or a combination of the models discussed earlier in the text.

10. In this particular greenhouse for which my guide gave me the data presented in the chapter, some products (e.g., mushrooms) were also cultivated in dark rooms with artificial light—which could explain the particularly large electricity bill.

11. In 2012, there were still 1786 state farms, producing 33.71 million tons of grain on 4.726 million hectares (National Bureau of Statistics).

12. The enterprise had told me that they had not yet started to harvest their own plot because it was too soon and the fruits were not ripe. However, oranges were already being packed and sent to Z.

13. Original language: "新中国成立以来,事实已经证明任何敌人都不可能用武力征服我们。然而,那种杀人不见血的生物武器则有可能使我们丧失警惕。" (*Xin zhongguo chengli yilai, shishi yijing zhengming renhe di rend ou bu. keneng yong wuli zhengfu women. Ran'er, na zhong sharen bujian xie de shengwu wuqi ze you keneng shi women sangshi jingto*).

14. Original language: "Ces nouveaux acteurs ont la capacité d'édicter de façon autonome des ensembles cohérents de règles, ayant vocation à s'imposer à un maximum de producteurs, sinon à leur totalité, qui deviennent parfois des référents pour l'action publique, et que l'on peut donc considérer comme des formes de politiques *publiques privées*."

References

ASTI Database. http://www.asti.cgiar.org/data/

Balme, R., & Tang, R. (2014). Environmental governance in the People's Republic of China: The political economy of growth, collective action and policy developments—Introductory perspectives. *Asia Pacific Journal of Public Administration, 36*(3), 167–172. https://doi.org/10.1080/23276665.2014.942067.

Botha, C. (1999). From mercenaries to "private military companies": The collapse of the African state and the outsourcing of state security. *South African Yearbook of International Law, 24*, 133–148.

Burns, J. P., Peters, B. G., Wang, X, & Li, J. (2010, June 17–19). *2010 Food safety policy coordination in three Chinese cities*. Paper prepared for the conference on 'Regulation in the age of crisis'. Third Biennial Conference of the Standing Group on Regulatory Governance of the European Consortium for Political Research ECPR and the Regulation Network, University College Dublin.

Conway, G. (2000, March 28). *Crop biotechnology: Benefits, risks and ownership.* Speech delivered by the president of the Rockefeller Foundation delivered at the OECD Edinburgh conference on the scientific and health aspects of genetically modified foods.

Crozier, M., & Friedberg, E. (1977). *L'Acteur et le Système.* Paris: Editions du Seuil.

De Janvry, A., Sadoulet, E., & Zhu N. (2005). *The role of non-farm incomes in reducing rural poverty and inequality in China.* (Working paper series). Berkeley: Department of Agricultural & Resource Economics, UC Berkeley.

Duchatel, M. (2011). Comment éviter de nouveaux scandales alimentaires? *China Analysis, 33,* 26–28.

Economy, E. C. (2007). The great leap backward? The costs of China's environmental crisis. *Foreign Affairs, 86*(5), 38–59.

ETC Group. (2008). *Who owns nature? Corporate power and the final frontier in the commodification of life.* Ottawa: ETC Group.

FAO. (2004). *The state of food and agriculture 2003–2004. Agricultural biotechnology meeting the needs of the poor?* Rome: FAO.

FAO. (2008). *An introduction to the basic concepts of food security.* Rome: EC—FAO Food Security Programme.

FAO. (2011). *FAO at work 2010–2011: Women key to food security.* Rome: FAO.

FAO Database. http://faostat3.fao.org/faostat-gateway/go/to/home/E

Fouilleux, E. (2010). Standards volontaires: entre internationalisation et privatisation des politiques agricoles. In B. Hervieu, N. Mayer, P. Muller, F. Purseigle, & J. Remy (Eds.), *Les mondes agricoles en politique* (pp. 371–396). Paris: Presses de Sciences Po.

Froissart, C. (2007). *Quelle citoyenneté pour les travailleurs migrants en République Populaire de Chine?: l'expérience de Chengdu.* Thèse: Sciences Politiques: Paris: Institut d'Etudes Politiques.

Gale, F. (2013). *U.S. exports surge as China supports agricultural prices.* USDA ERS. http://www.ers.usda.gov/amber-waves/2013-october/us-exports-surge-as-china-supports-agricultural-prices.aspx#.VIbMpdKG8wc

Goulet, F. (2010). Chapitre 1: Nature et ré-enchantement du monde. In B. Hervieu, N. Mayer, P. Muller, F. Purseigle, & J. Remy (Eds.), *Les mondes agricoles en politique* (pp. 51–72). Paris: Presses de Sciences Po.

Greenpeace. (2013). *Chinese herbs: Elixir of health or pesticides cocktail?.* http://www.greenpeace.org/international/Global/eastasia/publications/reports/food-agriculture/2013/chinese-herbs-pesticides-report.pdf

Han, Z. (2003). *De l'autonomie des entreprises d'Etat en droit chinois: le « gradualisme » de la réforme chinoise.* Paris: Budapest; Torino: L'Harmattan.

Heffernan, W. (1999). *Consolidation in the food and agriculture system.* Columbia: Department of Rural Sociology, University of Missouri.

Hibou, B. (1998). Retrait ou redéploiement de l'Etat? *Critique internationale, 1,* 151–168.

Ho, P. (2007). Embedded activism and political change in a semi authoritarian context. *China Information, 21*(2), 187–209.

Hu, D. (2013, November 13). *The opportunity & challenges of farmer-supermarket direct purchase in China.* Paper presented at the FAO's policy forum on rural-urban income gaps and smallholder market integration in Asia, Beijing.

Ikenberry, G. J. (2003). Conclusion. In T. V. Paul, J. A. Hall, & G. J. Ikenberry (Eds.), *The nation-state in question.* Princeton: Princeton University Press.

Jahiel, A. R. (1998). The organization of environmental protection in China. *The China Quarterly, 156,* 757.

Jin, X., Xu, Q., & Huang, C. (2005). Current status and future tendency of lake eutrophication in China. *Science in China Series C: Life Sciences, 48*(2), 948–954.

Jua, X., Xing, G., Chen, X., Zhang, S., Zhang, L., Liu, X., Cui, Z., Yin, B., Christie, P., Zhu, Z., & Zhang, F. (2009). Reducing environmental risk by improving N management in intensive Chinese agricultural systems. *PNAS, 106*(9), 3041–3046. https://doi.org/10.1073/pnas.0813417106.

Knoke, D. (1996). *Comparing policy networks: Labor politics in the US, Germany and Japan.* Cambridge: Cambridge University Press.

Kohut, A., & Wike, R. (2013). *Environmental concerns on the rise in China.* Pew Research Center. http://www.pewglobal.org/files/2013/09/Pew-Global-Attitudes-Project-China-Report-FINAL-9-19-132.pdf

Kung, J. (1999). The evolution of property rights in village enterprises. In J. C. Oi & A. Walder (Eds.), *Property rights and economic reform in China* (pp. 95–122). Stanford: Stanford University Press.

Kung, J. K. S., & Cai, Y. (2000). Property rights and fertilizing practices in rural China: Evidence from Northern Jiangsu. *Modern China, 26*(3), 276–308.

Li, H., & Rozelle, S. (2003). Privatizing rural China: Insider privatization, innovative contracts and the performance of township enterprises. *The China Quarterly, 176,* 981–1005.

Liu, W., & Zhang, W. (2011). Academician of the Chinese academy of engineering: 300 million mu of arable land are polluted by heavy metals. *Yangcheng Evening News 12 Oct 2011* [刘玮宁, 张炜哲, 工程院士称全国3亿亩耕地受到重金属污染, 羊城晚报, Liu Weining, Zhang Weizhi, Gongcheng yuanshi cheng quanguo 3 yi mu gengdi shoudao zhongjinshu wuran. *Yangcheng wanbao*]. http://www.chinanews.com/gn/2011/10-12/3383763.shtml. Accessed 4 Mar 2014.

Marin, B., & Mayntz, R. (1991). *Policy networks: Empirical evidence and theoretical considerations.* Boulder/Frankfurt am Main: Westview/Campus Verlag.

Marsh, D., & Rhodes, R. A. W. (1992). *Policy networks in British government.* Oxford: Clarendon Press.

Ministry of Agriculture. (2013, August 31). *GM and non-GM food are similarly safe.* News office of the ministry of agriculture [«转基因食品与非转基因食品具有同样的安全性 », 农业部新闻办公室]. http://www.moa.gov.cn/zwllm/zwdt/201308/t20130831_3592472.htm

National Bureau of Statistics Database. http://data.stats.gov.cn/workspace/index?m=hgnd

Nee, V., & Opper, S. (2012). *Capitalism from below: Markets and institutional change in China*. Cambridge, MA/London: Harvard University Press.

Nguyen G., & Purseigle F. (2012). The emergence of "firm" agriculture in France: Characteristics and coexistence with family farms. Paper presented at the IFSA symposium 2012, Workshop 1.3.

Oi, J. C. (1992). Fiscal reform and the economic foundations of local state corporatism in China. *World Politics, 45*(1), 99–126. https://doi.org/10.2307/2010520.

Oi, J. C. (1999a). *Rural China takes off: The institutional foundations of economic reform*. Berkeley: University of California Press.

Oi, J. C. (1999b). Two decades of rural reform in China: An overview and assessment. *The China Quarterly, 159*, 616–628.

Oi, J. C., Singer Barbiaz, K., Zhang, L., Luo, R., & Rozelle, S. (2012). Shifting fiscal control to limit cadre power in China's townships and villages. *The China Quarterly, 211*, 649–675.

Pei, X. (2002). The contribution of collective landownership to China's economic transition and rural industrialization: A resource allocation model. *Modern China, 28*(3), 279–314.

Peng G. (2013, August 21). Expert asks about GM staple grain: Why does China want to introduce unnecessary things? *Global Times* [彭光谦, «专家八问主粮转基因化:我国究竟为何要盲目引进», 环球时报]. http://finance.huanqiu.com/comment/2013-08/4267575.html

Peters, G. B. (1993). Managing the hollow state. In K. Eliassen & J. Kooiman (Eds.), *Managing public organizations: Lessons from contemporary European experience* (pp. 46–57). London: Sage.

Peters, G. B., & Pierre, J. (1998). Governance without government? Rethinking public administration. *Journal of Public Administration Research and Theory, 8*(2), 223–243. https://doi.org/10.1093/oxfordjournals.jpart.a024379.

Purseigle, F. (2012). Introduction. *Etudes Rurales, 190*(2), 19–23.

Rhodes, R. A. W. (1994). The hollowing out of the state: The changing nature of the public service in Britain. *Political Quarterly, 65*(2), 138–151.

Richardson, J. J., & Jordan, A. G. (1979). *Governing under pressure: The policy process in a post-parliamentary democracy*. Oxford: Robertson.

Rozelle, S., Huang, J., & Otsuka, K. (2005). The engines of a viable agriculture: Advances in biotechnology, market accessibility and land rentals in rural China. *The China Journal, 53*, 81–111.

Schmitter, P. C. (1974). Still the century of corporatism? *The Review of Politics, 36*(1), 85–131.

Schwoob, M. H. (2013a). L'éveil vert de la société chinoise? *Ecologie & Politique, 47*, 27–37.

Schwoob, M. H. (2013b). La réforme de la finance rurale. *China Analysis, 46*, 38–42.

Tang, M. (2012, November 11). Allowing rural finance to be more inclusive. *Caijing Magazine* [汤敏, "让农村金融更普惠", 《财经》杂志. Tang Min, Rang nongcun jinrong geng puhui. *Caijing Zazhi*].

Thornton, P. M. (2009). University crisis and governance: SARS and the resilience of the Chinese body politic. *The China Journal, 61*, 23–48. https://doi. org/10.1086/tcj.61.20648044.

Tilt, B. (2013). Industrial pollution and environmental health in rural China: Risk, uncertainty and individualization. *The China Quarterly, 214*, 283–301.

US-China Business Council. (2007). *Food safety and inspection in China.* Washington/Beijing/Shanghai: US-China Business Council.

Unger, J., & Chan, A. (1996). Corporatism in China: A developmental state in an East Asian context. In B. L. McCormick & J. Unger (Eds.), *China after socialism: In the footsteps of Eastern Europe or East Asia?* Armonk: Sharpe.

USDA. (2009, December 14). *Fertilizer—China. GAIN report CH9082,* Washington, DC: USDA Foreign Agricultural Service.

USDA. (2012, July 13). Agricultural biotechnology annual 2012. *GAIN report CH12046,* Washington, DC: USDA Foreign Agricultural Service.

Wang, J. (2005). Going beyond township and village enterprises in rural China. *Journal of Contemporary China, 14*(42), 177–187. https://doi.org/10.1080/10670560420003041 56.

Wang, P. (2012, November 11). Breaking and building rural finance. *Caijing Magazine* [王培成, "农村金融破与立", 《财经》杂志. Wang Peicheng, Nongcun jinrong po yu yi. *Caijing Zazhi*].

Whiting, S. H. (2000). *Power and wealth in rural China: The political economy of institutional change.* Cambridge: Cambridge University Press.

Wilson, F. L. (2012). *Interest-group politics in France.* Cambridge: Cambridge University Press.

Yan, T. (1996). The culture of Guanxi in a North China village. *The China Journal, 35*, 1–25.

Zhang, F. (2010, September). *Reforming China's state-owned farms: State farms in agrarian transition.* Paper presented at the 4th Asian Rural Sociology Association International Conference at Legazpi City, Philippines.

Zhang, X., Huang, J., Qiu, H., & Huang, Z. (2010). A consumer segmentation study with regards to genetically modified food in urban China. *Food Policy, 35*, 456–462.

Zhang, F., Cooke, P., & Wu, F. (2011). State-sponsored research and development: A case study of China's biotechnology. *Regional Studies, 45*(5), 575–595.

Zhou, L., & Takeuchi, H. (2010). Informal lenders and rural finance in China: A report from the field. *Modern China, 36*(3), 302–328.

Zhou, X., Lian, H., Ortolano, L., & Ye, Y. (2013). A behavioral model of "muddling through" in the Chinese bureaucracy: The case of environmental protection. *The China Journal, 70*, 120–147.

CHAPTER 4

The Grip of Local States

4.1 Local Control Mechanisms

All in all, whether in making investments themselves or in regulating businesses, the conventional wisdom is that agents of the Chinese state tend to exercise power arbitrarily, often in search of rents individually or institutionally. (Yang Dali, *Remaking the Chinese Leviathan*)

4.1.1 Institutional Fragmentation and the Power of Local States

What is usually called "the state" in China is in fact heavily fragmented and made of an array of players, from central to local levels. Post-Maoist decentralization reforms gave considerable power to local officials. The fiscal system, in particular, underwent consequent changes. Whereas under the Maoist era, local governments were not granted with any decision making power in terms of public expenditures (consolidated budgets were fixed by the central level, which then ratified local budgets according to their estimated needs), the 1980s saw the establishment of three different types of revenues: central-fixed revenues, local-fixed revenues and shared revenues. Local bureaus became the only institutional entities responsible of collecting taxes. This greatly increased the power of local officials, who took advantage of the situation and started establishing a network of close ties with local enterprises, from which they were collecting taxes. At the

© The Author(s) 2018
M.-H. Schwoob, *Food Security and the Modernisation Pathway in China*, Critical Studies of the Asia-Pacific,
https://doi.org/10.1007/978-3-319-65702-8_4

beginning of the 1990s, central revenues started shrinking at a rapid pace, resulting in a fiscal stress that pushed the central government to take measures to restore its control over the fiscal system. In 1994, national tax bureaus were created and clear shares for national and local revenues were established. In spite of these attempts of recentralization, the share of revenue collected by local governments as well as their share in government spending (two figures that are commonly used to evaluate the degree of decentralization of a given country) kept on rising. The share of expenditures of local governments was almost 75 percent in 2005 (compared to 19.6 percent in developing economies and 32 percent in OCDE countries), whereas their share in the national revenue was 48 percent (compared to 19.6 percent in developing economies and 32 percent in OCDE countries) (Shen et al. 2012: 3).

At the local level, the fragmentation of governmental bodies in charge of agricultural policies is similar to what can be observed at the central level. Administrative units are organized according to a hierarchy ranging from the most central institutions (e.g., the NDRC, the State Council, and the ministries) to the most local bodies (provinces, municipalities, and autonomous regions; prefectures; districts and countries; towns; villages). In the course of the policy-making process, highest government institutions take the most general decisions, which are then progressively detailed among their descent in the lowest ranks of administrative bodies. Policy implementation is thus sequentially and geographically fragmented (Lieberthal 1992). Local governments operate inside "branches" (条 *tiao*), formed by vertical center-periphery hierarchy. Agricultural bureaus, for instance, work under the supervision of the MOA. In addition, local governments operate inside horizontal "lumps" (块 *kuai*) as well, which are local bureaus. This aspect greatly complicates the political process at the local level, as Lieberthal (1997: 3) explains: "One key rule of the Chinese system is that *units of the same rank cannot issue binding orders to each other.* [...] The natural consequence of this operating rule is that there often is a tremendous need to build a consensus in order to operate effectively in China, and negotiations aimed at consensus building are a core feature of this system."

At the local level, the *tiao tiao kuai kuai* structure (Table 4.1) makes bargaining unavoidable. According to Lampton (1992), although bargaining already existed prior to 1978 (administrations were already organized according to territorial levels), post-Maoist reforms further amplified their importance. Decentralization of economic power indeed made the

Table 4.1 Hierarchical structure of the planning and executive branches linked to agricultural and rural policies in China

National State Council

National Development and Reform Commission (former State Planning Commission)	Ministry of Agriculture	State-owned Assets Supervision and Administration Commission (SASAC) SOEs (Sinograin, COFCO)	General Administrations: General Administration of Quality Supervision, Inspection and Quarantine, General Administration of Customs…	Other Ministries: Ministry of Foreign Affairs, Ministry of Education, Ministry of Commerce, Ministry of Water Resources, Ministry of Land and Resources, Ministry of Environmental Protection, Ministry of Housing and Rural-Urban Development, Ministry of Health (and State Food and Drug Administration directly under its supervision)…
Department of Rural Economy … State Grain Administration				
	Local bureaus (province, prefecture, county, township, and village levels)	Local branches	Entry-Exit Inspection and Quarantine Bureaus and Bureaus of Quality and Technical Supervision	Local bureaus (province, prefecture, county, township, and village levels)

number of local organizational bases proliferate and increased their power. In policy formulation, central bureaucracies have to bargain with empowered territorial administrations. In the course of policy implementation, the highest bodies of the government have to negotiate with stronger subordinated bureaus to gain their support and ensure their cooperation.

At the central level, legislative processes are "frequently unable or unwilling to arrive at precise settlements of the conflicting interests on many issues. Only by leaving some matters somewhat nebulous and unsettled can agreement on legislation be reached" (Anderson 2003). A fundamental consequence of such a fragmentation of the political process is that local officials have important decision-making power in the carrying out of policies. In particular, county-level officials have decision-making power in the carrying out of agricultural modernization policies. However, all the county-level bureaus do not have equal power in the process and the power of a given bureau greatly varies from one place to another. Bargaining indeed does not only take place vertically but horizontally as well. Inside local "lumps", bureaus have to negotiate with each other. For instance, local agricultural bureaus are responsible of allocating agricultural subsidies, but also depend on local financial bureaus to have access to public funds.

One of the most striking things I could acknowledge when conducting fieldwork was the wide variety of local bodies in charge of implementing agricultural policies, which varied from one place to another. I was often regarded as a foreign investor—or, at least, as a foreigner able to provide resources (financial resources, professional or political contacts, expertise, etc.) to contribute to agricultural development in the area I was visiting. As such, I was directed toward local bureaus in charge of cooperation, investment, and agricultural development. In the various places I went to, my main interlocutor varied greatly: while in some places, it seemed that the investment promotion bureau was the most important local institution in charge of developing agricultural projects, in others, the role was rather taken by the poverty alleviation bureau, or the sustainable rural development "association" (*xiehui* 协会), the fruit development bureau, the agricultural development bureau, the grain bureau, and so forth.

What was striking was that local institutions I used to think would naturally cooperate with each other (e.g., the poverty alleviation bureau with the agricultural development bureau in low-income rural areas) were not necessarily working together and sometimes even barely knew each other. Interviews conducted in a county of Chongqing demonstrated that the poverty alleviation bureau was willing to take credit for poverty alleviation achievements, without including the agricultural development bureau, as officials of the two bodies were in fact competing against each other in their struggle for higher positions in the local political hierarchy.

This does not mean that cooperation never happens between local bureaus. In Lanshui county in the Shandong province, I met government officials from the investment promotion bureau who were working with the fruits development bureau. In fact, local officials can take the decision to cooperate with each other when both parties *wish to achieve the same results* (i.e., have the same goals and interests) and think that cooperation *will not jeopardize their careers' progress.* This is a plausible explanation for the cooperation between the investment promotion bureau and the fruits development bureau in Shandong. People of the two institutions indeed have very different profiles and career opportunities. People working at the investment promotion bureau were mostly government officials, whereas researchers formed the majority of the employees of the fruit development bureau.

Just like I did in the course of my fieldwork, entrepreneurs willing to launch business in food-processing or to establish direct links with rural producers interact with different local bureaus, depending on established local patterns of power and on networking opportunities they are able to grasp. The leeway granted to local bureaus for the implementation of agricultural investment promotion policies and the lack of rules clearly establishing the responsibilities of each bureau in the process are important factors that increase the power of local governments over enterprises. This power is exercised through a variety of domination mechanisms.

4.1.2   *Financial Resources: Subsidy Mechanisms*

The analysis of local patterns of power of Chap. 3 showed that one of the most important upstream environments of agricultural production was access to finance, and that whereas farmers were strongly suffering from a lack of access to credit, enterprises were benefitting from a much higher degree of trust in the banking sector. Local governments, on their side (at least, at the county-level and above), also have access to important financial resources, especially since agriculture and rural areas have been prioritized by the central government. In the third chapter, constraints in terms of *access to credit* were depicted, but the issue of *access to governmental subsidies* was not described in detail. According to interviews conducted during fieldwork, access to governmental subsidies is a strong incentive for entrepreneurs to engage in the agricultural business. Agriculture indeed usually generates few profits and has low return on investment compared to other sectors of the economy. As a consequence, entrepreneurs have little control over this uncertainty—access to subsidies—whereas, on the opposite, local governments have important leeway on the decision to attribute subsidy policies, which allows them to gain a significant advantage over rural enterprises.

An interesting observation gained from fieldwork is that agricultural subsidies vary greatly depending on areas. Subsidies differ in three ways: in the range of products covered by the local scheme, in the amount of subsidies given per unit of product/per hectare, and so on, and in the allocation process (who receives the subsidy: the buyer or the producer; is it a subsidy per hectare or per unit of product; is it directly transferred on bank accounts or do people receive the subsidy in another way; what conditions have to be fulfilled by people applying for subsidies, etc.). To illustrate these differences, Table 4.2 provides details on the agricultural machinery subsidy systems of Huangmo (Ningxia), Lushan (Jiangxi), and Lanshui (Shandong). Agricultural machinery subsidies are among the most widespread types of agricultural subsidy throughout China, compared, for instance, to grain subsidies that essentially target grain-producing areas. Table 4.2 provides extracts (translated and sometimes summarized) from three documents: (i) "Huangmo's 2012 procedure to handle subsidies for agricultural machinery purchase" (published at the end of the year 2012); (ii) "Lushan's reform of the subsidy procedure for the purchase of agricultural machinery" (published at the beginning of the year 2013); (iii) "Lanshui's 2013 first batch of policies linked to subsidies for the purchase of agricultural machinery" (published in June 2013). As Table 4.2 illustrates, procedures vary greatly from one area to another, as well as the list of subsidized products and corresponding amounts, and some areas even have lists of "approved enterprises" for agricultural machinery subsidies. The lack of standardized procedures and the sometimes very complicated steps that need to be taken to have access to subsidies is another factor that empowers local bureaus.

Even though procedures differ greatly among provinces, local rules usually have in common that farmers and agricultural machinery manufacturers are their sole beneficiaries. On the opposite, food enterprises I met during my fieldwork complained that they did not have access to subsidies

Table 4.2 Agricultural machinery official procedures in Huangmo, Lushan, and Lanshui

County	Huangmo	Lushan	Lanshui
Issuance office	The document was issued by "the township agricultural machinery departments" (乡(镇)农机部门 *xiang (zhen) nongji bumen*), drafted according to the document issued by Ningxia bureau of agricultural machinery (农机局 *nongjiju*) (宁夏农业机械购置补贴办理流程 [Ningxia procedure for subsidies for the purchase of agricultural machinery]) and adapted to the "local situation."		The document was issued by the city government, drafted according to the requirements of the document issued by Yantai government (关于印发2013年烟台市农业机械购置补贴工作实施方案的通知 [Information on Yantai's implementation program of subsidies for the purchase of agricultural machinery]) and adapted to the "local situation."
Procedure	Farmers' households should apply for agricultural machinery subsidies to the department in charge of agricultural machinery subsidies (乡(镇)农机部门 *xiang (zhen) nongji bumen*) of their township, which will weekly hand over reports to the county agricultural machinery extension services center (县农机推广服务中心 *xian nongji tuiguan fuwu zhongxin*). After examination of all information (name, models of machines and tools, quantity, phone number, ID card, amount of subsidy), a notice will be posted to the village committee, which will make it public. If there is no objection within seven days, the township cadres (乡(镇)主管领导和农机乡干 *xiang (zhen) zhuguan lingdao he nongji zhuangan*) will sign the agreement and the county extension services center will be able to start subsidy procedures.	People willing to buy agricultural machinery should go to the county or township-level departments in charge of agricultural machinery (县或乡镇农机部门 [*xian huo xiangzhen nongji bumen*) to apply for machinery and to the county-level departments in charge of agricultural machinery (县农机部门 *xian nongji bumen*) to get the notice to apply for subsidies.	The document states that applicants should be farmers (or agricultural workers: "农民(农场职工)" *nongmin (nongchange zhigong)*) or organizations directly engaged in the production of agricultural machinery. In order to apply for subsidies, farmers should come register themselves with their ID at Lanshui office for the management of agricultural machinery (农业机械管理办公室, *nongye jixie guanli bangongshi*). Subsidies are set according to a list established by the provincial government (山东省2012年农机购置补贴机具补贴额一览表 *shandongsheng 2012 nian nongji gouzhi butie jiju butie e yilanbiao*).

(continued)

Table 4.2 (continued)

County	Huangmo	Lushan	Lanshui
	Farmers should go to the county office for agricultural machinery subsidy (县农机购机补贴办公室 *xian nongji gouzhi butie bangongshi*) in order to fill in an application form and be provided with a notice. They then have a certain amount of time to go with the notice and their ID cards to a machinery broker, negotiate the price by themselves and purchase the machinery at a price decreased by the amount of subsidy. The sale is double-checked both by the county office for agricultural machinery subsidy (县农机购机补贴办公室 *xian nongji gouzhi butie bangongshi*) and the county financial bureau, and the machinery broker gets the subsidy.	Then, this people should select a broker in the province within ten days, buy agricultural machinery, and apply for subsidies within three months to apply for subsidy. After a multiple-steps checking process (by the county agricultural machinery department (县农机部门 *xian nongji bumen*), the county financial bureau, the township agricultural machinery department (乡镇农机部门 *xiangzhen nongji bumen*) and the township financial bureau), the subsidy is finally granted.	
Special conditions	Machinery cannot be resold within two years.	Subsidy per machinery cannot exceed 50,000 RMB.	Machinery cannot be resold within two years.

for agricultural inputs or machinery, even when they grew their own crops. However, fieldwork demonstrated that the complexity of procedures was difficult to overreach for farmers. The lack of information, the scarcity of vehicles, local language barriers, and low education level were named as major roadblocks preventing farmers to have access to subsidies. As a consequence, what I usually saw—in the case of Lushan and Lanshui—was that food enterprises were helping farmers to get subsidies, either by applying for them or by creating farmers' cooperatives, in the name of which the enterprise would then buy agricultural machinery for farmers-employees.

Procedures can reach a degree of complexity so high that trainings to get subsidies are sometimes provided by government officials. What is interesting is that enterprises attend such trainings. On 17 February 2014, the agricultural machinery bureau of Lushan invited the local financial bureau, local media, rural credit cooperatives as well as agricultural machinery manufacturers to attend trainings on subsidies. Farmers were not mentioned in the list of trainees. The director of the financial bureau said that he was "hoping that agricultural machinery enterprises would disseminate [the information across rural areas]".[1] Here again, we see another demonstration of how government officials see local enterprises as multipliers for agricultural modernization.

In the earlier discussion, I chose to depict the case of agricultural machinery subsidies because it is the most widespread agricultural subsidy scheme in the country. As such, official documents were easier to find and to compare. However, the same remarks apply to other types of subsidies as well, such as the ones for farm equipment (greenhouses, pest traps, etc.) or for seeds, pesticides, or fertilizers. The only kind of agricultural subsidy for which local officials probably play a less important role is per-hectare subsidies. Although they still constitute a small share of the whole agricultural subsidy scheme, direct subsidies have been developing quickly in the past few years. They usually work with a bank account system, funds being directly transferred on farmers' bank accounts each year according to the size of the cultivated area. However, per-hectare subsidies also vary greatly from province to province, because they depend on the funds granted by central administrations to local grain bureaus.

The haziness and complexity of procedures and the fact that enterprises are better equipped than farmers to overcome these obstacles, but are usually not able to get subsidies directly, create considerable advantages for local governments over rural enterprises. County (县 *xian*) and township

(乡 *xiang*) governments, in particular, gain significant power in the process—even if they remain highly dependent of decisions made by higher governmental bodies, both for the allocation of funds and for the evolution of their careers. Farmers, on their side, can have access to information regarding the organizational rules of the local subsidy scheme only through village and township-level government institutions. The following quote of an interview conducted in Jiangxi with a manager of an orange processing factory well illustrates these conclusions:

> We need to build a warehouse, and for that we need to apply for subsidies to the fruit industry bureau, which is a department in charge of managing the fruit sector. Some subsidies are granted by the fruit industry bureau, others are granted by other governmental bureaus [...] We need the support of the government, we must have it. (Interview, Jiangxi, October 2013)[2]

The agricultural subsidy scheme is on its way of getting simpler. In Lushan, for instance, the government recently expressed its wish to establish a shorter four-step procedure for agricultural machinery subsidies: in the future, farmers shall buy machinery and apply for subsidies, before governments check machinery and allocate subsidies. The procedure seems also on its way to becoming more transparent in Lushan, where the government is willing to establish an information disclosure system (农机购置补贴信息公开制度 *nongji gouzhi bujie xinxi gongkai zhidu*), a responsibility system (农机购置补贴工作责任制度 *nongji gouzhi butie gongzuo zeren zhidu*) and a complaint management system (农机购置补贴信访投诉管理制度 *nongji gouzhi butie xinfang tousu guanli zhidu*). However, at the time fieldwork was conducted, the complexity of procedures still granted local officials with important power over entrepreneurs.

4.1.3   Control Over Nonfinancial Resources

Local subsidy schemes are essential to agricultural investors. However, perhaps more importantly, local governments also provide nonfinancial resources that are vital to agribusiness, such as land, human resources, and certificates. The ability to control these upstream environments, again, grants local governments with an important power over rural enterprises.

In spite of China's rapid urbanization process, which could have freed the agricultural sector from labor surpluses and enabled farmers to cultivate bigger farms, data show that the fall in the number of farmers employed by the agricultural sector (a consequence of urbanization) did not much influence the size of cultivated land per capita, which remained stable (around 2 mu) after a jump at the beginning of the 1980s caused by agricultural reforms (World Bank database and National Bureau of Statistics).

Enterprises willing to invest in agricultural production have two options regarding arable land: they can either rent plots by themselves or contract with farmers. In both cases, entrepreneurs need to get the agreement of an important number of small farmers, a part of which does not live in rural areas anymore. In order to make things easier, entrepreneurs usually choose to address county and township governments to "organize farmers", especially when they come from another area of the country. Particularly enlightening was this sentence from an agribusinesswoman established in Beijing:

> [The project in Beijing] has become a pilot project. [...] Now, we are starting to launch projects in other provinces. Local governments come to look for us. [...] They organize land and farmers and have them ready for us. (Interview, Beijing, June 2013)

In addition, even if enterprises theoretically need to get the agreement of farmers, land *property* remains in the hands of the government, which considerably increases the capacity of local governments to "organize land" or to "organize farmers"—usually through the involvement of county, township, and village-level government officials. Sometimes, land is allocated to enterprises without getting the agreement of farmers. This happens, for instance, when virgin land (unoccupied and uncultivated land) is converted into land suitable for farming. In Jiangxi, many of the entrepreneurs I met had started business at the beginning of the 2000s, when they were offered the opportunity to plant citrus orchards on hills formerly covered with forests (Figs. 4.1 and 4.2). Orchards, in this area, are indeed considered as "forests". Converting hills into orchards is easy, as it does not change the land classification. In addition, orchards are particularly advantageous for enterprises, as forest lease contracts last longer than farmland lease contracts. In places I

Fig. 4.1 Certificates for forest rights granted to enterprises at the beginning of the 2000s (林权证 *lin quan zheng*)

went to in Jiangxi, this period usually went up to 50 years. In such cases, local governments of the county or township level become unavoidable negotiating partners.

Apart from these exceptional cases where virgin land is converted into land suitable for agricultural production, the degree of difficulty to "organize farmers" usually depends on the degree of industrial development of the area. In industrialized regions—or in rural areas close or well connected to industrial regions—farmers have greater opportunities to find jobs outside the agricultural sector. As a consequence, it is usually easier to get land from farmers in these areas. As was explaining a manager at X. conducting projects in Shandong and Jiangxi:

> In Shandong, it is easier to gather land to create big farms, because in this area, farmers go to cities. Sometimes, they rent their land to other farmers, sometimes they give it because they just don't care, they have better lives in cities. Here [in Jiangxi], it is more difficult to gather land, because farmers don't have any other source of income. (Interview, Jiangxi, October 2012)

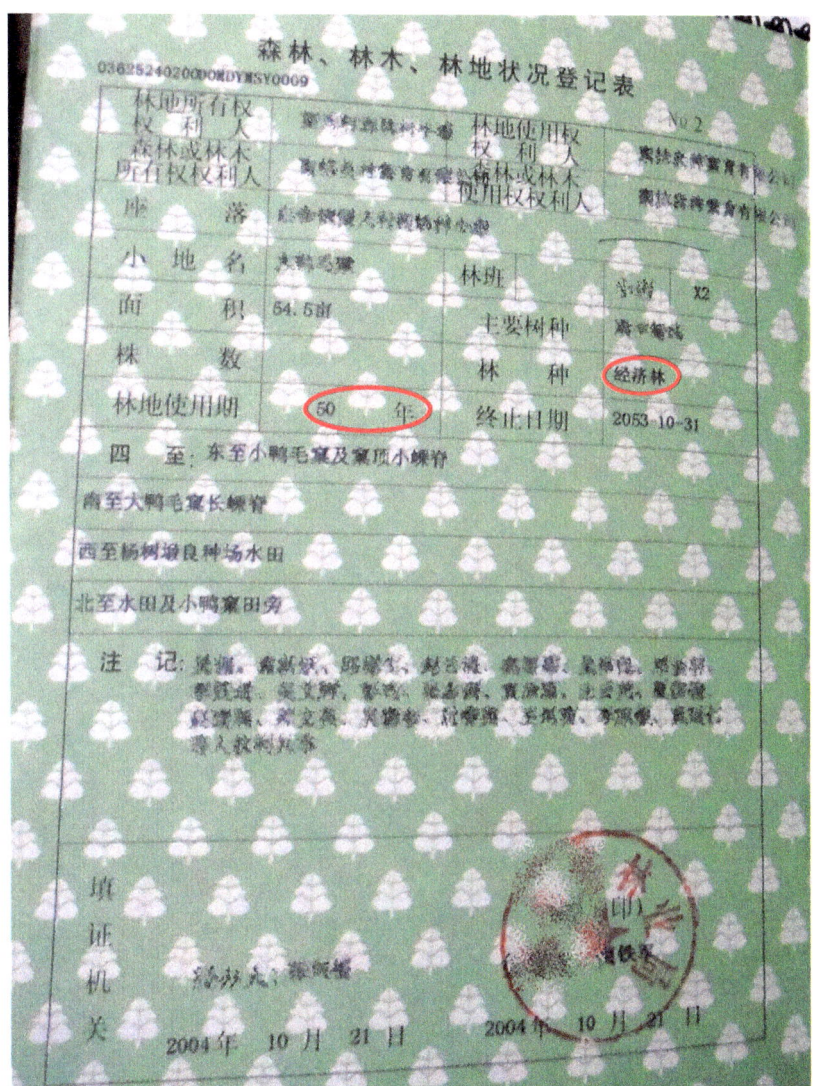

Fig. 4.2 Inside page of one certificate (*Circled in red*: "林种: 经济林" *linzhong: jingji lin* [type of forest: economic forest] and "林地使用期: 50年" *lindi shiyong qi: 50 nian* [duration of use: 50 years])

In addition, the degree of difficulty of gathering land also depends on the origin of the entrepreneurs. For enterprises founded and managed by local people, it is usually much easier to find farmers without the help of the government. As a manager of a rural food enterprise founded by a farmer in Jiangxi was telling me:

> Because everyone is local people (本地人 *bendiren*), we have a clear view of the situation of scattered [rural] households. People all know each other, [this is] Chinese kinship relations (中国人的亲戚关系 *zhongguoren de qinqi guanxi*). (Interview, Jiangxi, October 2013)

However, even local enterpreneurs need the support of the government. These latest indeed have to rent nonagricultural land to build plants, warehouses, or office buildings. In addition, they also need the support of the local government to create incentives so that farmers are encouraged to keep on farming. As was saying a manager of an orange factory in Jiangxi:

> We need the government to call on farmers to cultivate oranges. [We have to] support farmers by giving money for every tree they plant: for example, if a tree costs 3 RMB, the government will give 2 RMB, you [as a farmer] will pay only 1 RMB, this will encourage you to plant trees. (Interview, Jiangxi, October 2013)

Sometimes, local governments decide to create agricultural development zones. The control of the government over land and human resources, in these areas, is particularly strong. I had the opportunity to visit one near Changzhou, in Jiangsu province. The area was labeled "modern agriculture demonstration zone" (现代农业示范区 *xiandai nongye shifan qu*) and described by my guide (working at the grain bureau of Changzhou) as "an industrial development zone, but for agriculture" (Interview, Jiangsu, June 2013). The agricultural development zone was created in 2009 and divided in several subareas, in which investors could "make their choice". As was explaining one manager of the area:

> Usually, investors take areas of 3,000 mus. They rent land at 8-900 RMB per year. Leasing contracts last from 30 to 50 years. […] The principle of investments is as follows: investors arrive, rent land to peasants who move

from the status of peasant (农民 *nongmin*) to the status of workers (工人 *gongren*). (Interview, Jiangsu, June 2013)

To sum up, the fact that land legally belongs to the state and is fragmented in small plots rented by numerous and mobile farmers grants local governments—village committees, who have direct links with farmers, but mostly township and county governments, who have a strong capacity to influence lower levels and "organize" rural resources—with significant power over rural enterprises.

The control of reputational resources and access to market constitutes another factor increasing the power local officials have over rural enterprises. Local officials can indeed act as intermediaries between food-processing factories and potential buyers of agricultural products. Since 2007, an increasing number of retailers based in urban areas have been willing to purchase agricultural products directly from rural areas. Local governments sometimes intervene in the process. In the place I went to in Jiangxi, X., for instance, selected suppliers according to several lists: the one made by X.'s local sales managers, the one made by a Chinese professor hired by the company to assist the team in the development of DP, and the one made by the prefectural government. As was saying a manager in charge of looking for direct suppliers in rural areas:

Before, X. already had a long list of producers. [...] Today, we would like to expand the list. New producers are people who were recommended by the government or by professor [H.]. (Interview, Jiangxi, October 2013)

According to another manager:

It goes that way: the government tells us: you will work with this supplier, with this one here, with this one there. You will work with this slaughterhouse. It is all informal obligations of course. However, if we do not do it, we face the risk to find something [bad about us] in the media the week after. [...] In fact, nothing else is done outside of the government. (Interview, Shanghai, October 2012)

In addition to the pressure exerted on the headquarters of retail enterprises, pressure is also put on teams sent to the countryside to look for local suppliers. Government is omnipresent during visits. As was confessing a manager at X. conducting DP projects in Jiangxi:

Last time, we had dinner with the government. They didn't talk much about policies, they just said 'It's the best supplier, you should do business with him'. (Interview, Jiangxi, October 2012)

Townships to municipal governments have become intermediaries between producers and consumers and play a role in helping rural food enterprises to find new clients—another lever they can use to exert control on rural-based food-processing enterprises.

The last—but not least—kind of resource held by local officials is the ability to deliver licenses and certificates. Local bureaus are usually in charge of delivering licenses and certificates. Several kinds of business license are mandatory for food enterprises. The first license that they need is a food production license (食品生产许可 *shipin shengchan xuke*), granted by above-county-level Administrations of Quality and Technology Supervision. In addition, food enterprises need a healthy food production license (保健食品生产许可 *baojian shipin shengchan xuke*), delivered by the general Administration of Quality Supervision, Inspection and Quarantine (AQSIQ). Producers willing to sell products in other provinces also need a food circulation permit (食品流通 许可证 *shipin liutong xukezheng*), granted by the local bureaus of commerce. In addition to these mandatory licenses, enterprises might be willing to obtain documents such as the Global Gap certificate—this latest being quite popular in the areas where I conducted fieldwork. Quality certificates are granted by local certification bodies theoretically independent from the state but which still have to be approved by the PRC's Certification and Accreditation Administration (中国国 家认证认可监督管理委员会 *zhongguo guojia renzheng renke jiandu guanli weiyuanhui*), a body of the AQSIQ. A survey investigating certification agencies in Guangzhou, Shenzhen, Hangzhou, and Qingdao demonstrated that, in fact, "most [of them were] run by or affiliated with the government rather than being market-driven" (Fan et al. 2009: 628).

The fact that local governments are the sole players able to deliver mandatory licenses and licenses that are not mandatory but are necessary to develop business grants them with an important power over enterprises. In addition, enterprises are regularly checked and can see their license suspended or revoked. As a consequence, the granting of a given license does not put an end to the pressure exerted on entrepreneurs, who keep on making efforts to preserve it.

4.2   Entrepreneurs' Strategies and the Place of the State

The question surrounding the governance of China's markets, then, is not whether the government will remain involved but, rather, what form the new 'regulatory state' will take. (Margaret Pearson, *The Business of Governing Business in China*)

All the resources depicted earlier in the text—financial, material, reputational, and normative resources—grant local governments with an important capacity to control entrepreneurs (Fig. 4.3). Maintaining good relationships with the government is not just useful for agrifood-entrepreneurs to access crucial resources to start or expand their businesses. I was also explained that the government "could easily make things more difficult" to entrepreneurs through their capacity to grant, suspend, and revoke licenses and certificates, fundamental to trade food products in the current context of food safety issues. Regulations do exist but are used by local governments both in formal ways (through standard and institutionalized procedures) and in informal ways, which have to deal with the establishment of personal relationships and social networks, in which applicants are subjectively selected by local officials.

Interviews conducted in Jiangxi and Shandong showed that entrepreneurs were constantly worrying about maintaining good relationships with the government and continuously developing strategies aimed at fulfilling this goal. A variety of opinions were voiced by food enterprises concerning the action of local governments in rural areas. Some remarks expressed vehement criticism:

> [*Is it not the role of the government to spread agricultural techniques?*] The government doesn't care, just drinks wine! (Interview, Jiangxi, October 2012)

An equivalent number of remarks, on the opposite, expressed approbation:

> The government has policies to sustain the peasants, we need to build a warehouse and for that we apply for subsidies to the fruit industrial bureau. It will not grant them to you because you drank wine with them. China is a society ruled by law. (Interview, Jiangxi, October 2013)

Nevertheless, all the enterprises I met were highly valuing relationships with government officials—because of the resources they could provide

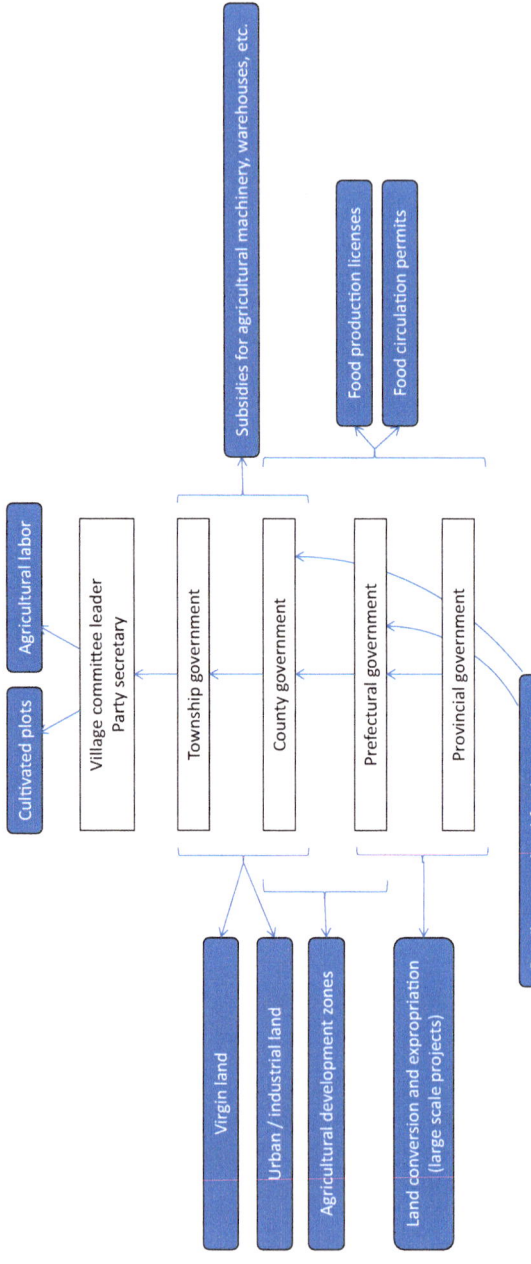

Fig. 4.3   Control of the different levels of local governments on financial and nonfinancial resources

and take away from them—and were developing strategies to establish and consolidate *guanxi*. Strategies could be rather simple and formal ones, taking place at the occasion of official visits, for instance, as the following quote from a manager of rural food processing enterprise illustrates:

> [We establish relationships with the government] because we have many things to deal with them, for example they pass by because they need to manage us. For example we have a meeting together, they also may come to our factory, they often come to check on our work. The government comes to see you in order to see if everything complies (有没有符合条件*you mei you fuhe jiaotian*). If I need to get money from the government, and if the government does not give me [money], they can come and check if you meet the standards of examination and approval. (Interview, Jiangxi, October 2013)

Strategies could be more creative and include the displaying of posters in offices and factories, presenting how enterprises were managed, how food safety was taken care of, and so on. What was striking in the messages displayed by such posters was that they seemed to essentially address governmental officials rather than potential customers, as the following extract illustrates:

> Each household or plot has a *danwei* [work unit] number, is drawn on a map and reported to the town and county-level fruit industry bureaus. (Translation from a poster displayed in a food processing enterprise, Jiangxi)

Strategies could also be way more elaborate and include social meetings such as dinners. In Jiangxi, an employee of a rural-based food enterprise told me, for instance:

> The manager has to leave us now because he has to have dinner with the financial department and the mayor. They have to pay some taxes and maybe after some drinks they will lower down the price. (Interview, Jiangxi, October 2012)

In another orange-processing enterprise:

> This is the factory's canteen. A lot of "*lingdao*" (领导 "leader") come to have dinner here. She says that it is good and comfortable, but also *guanxi* are essential most importantly. (Interview, Jiangxi, October 2012)

It often happened that managers could not receive me or had to leave at some point or the meeting for a while because they had to go to meetings with the government. "He is seeing leaders" (他现在看领导 *ta xianzai kan lingdao*) was the most widespread explanation that was given to me. The most "achieved" form of *guanxi* I saw was an entrepreneur who was friend with the mayor. During a visit of a rural-based orange factory in Jiangxi, the manager invited me to have lunch in a fancy restaurant in the township. We ran into the mayor, who invited us at his table. As a friend of him, the mayor was "often playing majong or poker" with the manager, and we had lunch altogether along with other local officials.

To sum up the discussion, the growing importance of food enterprises in the farming sector does not grant them with complete autonomy. Their new involvement in agricultural modernization over the last decade was indeed largely state-induced. By converting forests into farmland, by establishing agricultural investment zones or by implementing other incentive mechanisms, county, and township-level governments managed to attract investors since the beginning of the 2000s. In addition, in spite of the economic liberalization and the hollowing out of the state capacity to directly control agricultural production activities, local states managed to use their direct or indirect control of resources—such as land and human resources—and to develop regulatory and pseudo-regulatory control mechanisms—through their capacity to grant and withdraw subsidies and licenses. The existence of pseudo-regulatory mechanisms is permitted by the fact that regulations vary at the discretion of local governments, granted with significant leeway in the implementation of policies since decentralization.

The capacity of local officials to engage in economic networks and activities was described by a wide corpus of literature. Researchers depicted "developmental" (Blecher and Shue 1996) or "entrepreneurial" (Duckett 1996) states or portrayed forms of "local state corporatism" (Oi 1992). However, none of these theoretical frameworks really provided frames matching what I observed in Shandong and Jiangxi. The mechanisms described in Oi's model of local state corporatism are very similar to some of the mechanisms depicted earlier in the text. In Oi's model indeed, local governments keep control over enterprises through the contract responsibility system, through the allocation of key resources and through the providing of bureaucratic services, tax breaks, investment, and credit. However, as this present analysis remains limited to the agricultural sector, it seems rather difficult to argue that local states "act as business corporations" by taking profits from local factories "to pay for expenditures and for reinvestment," as it is the case in Oi's model. From what I could observe, local states indeed highly depend on above-level governments for agricultural subsidies, as agriculture itself does not generate profits for local states, since agricultural taxes were abolished in 2006. In addition, Oi's framework of local state corporatism does not provide answers to the following question: why did local governments decide to reinvest agricultural production activities, whereas these latest do not generate tax revenue that can be redistributed to other sectors?

In the areas where I conducted fieldwork, local officials were not taking over the role of entrepreneurs in food factories or retail enterprises. On the opposite, a clear distinction was always made between entrepreneurs and "*lingdao*"—a term always referring to government officials. As a consequence, the cases of agricultural production observed in Jiangxi and Shandong do not fit in the framework of entrepreneurial state either.

The theory of the developmental state is perhaps the most likely to suit the findings of this research. In this framework, enterprises (either state-, collective or private) undertake entrepreneurship under suitable conditions shaped by the government—among others, through the establishment of close relationships with selected business groups. Originally, the concept, framed by Chalmers Johnson, depicted developmental states as governments contributing to economic growth through the establishment of large national corporations controlled by dedicated ministries. Today, the theoretical framework has evolved a lot and refers to a broader notion according to which governments "dynamically help to create the political and infrastructural conditions for economic growth" (Blecher and Shue 1996: 109). However, even if the evolution of the concept enables the case studies of this research to fit in, the framework of developmental states sadly has lost a lot of its explanatory capacity. In addition, the framework was widely used to explain the role that East Asian states played in economic development, an approach that is too growth-centered. On the opposite, it is necessary for this research to "move beyond the growth perspective" (Boyd and Ngo 2005: 9), because agriculture plays a rather limited role in the national economic growth. For all these reasons, it is necessary to go further in the analysis.

Insights from fieldwork showed that more tribute had to be given to inherent social logics. To put it shortly, stakeholders need to be brought back in, in the whole complexity of their interactions, by adopting a relational approach to state capacity and power. The frameworks of developmental states and pseudo-regulatory states would gain a lot by being merged. In the course of economic liberalization, the planning functions of Chinese local states shrunk. In addition, their involvement in agricultural activities progressively hollowed out, as their interest shifted to industrial and urban development. Building on the impetus given by policy guidelines promulgated by the central state since the beginning of the 2000s, local state officials managed to reintegrate agricultural production activities through the development of ties and networks with private entrepreneurs (usually excluding farmers) and the reinvestigation of existing entrepreneurial networks. These latest were progressively used as tools for the coming back of developmental local states, which started to rely on resources at their disposal and on pseudo-regulations to control these networks—in the sense that they adapt loose regulations to the structure of social ties they build and maintain with entrepreneurs. To sum up, a transformation of existing regulations and of key resources into control tools helped local officials better control the developing network of food-processing enterprises in rural areas. These latest serve as transmission belts for agricultural modernization, allowing local governments to reinvestigate agricultural production activities.

Local state-enterprises networks recently evolved toward wider and more complex forms of social ties. Over the past few years, the multiplication of DP projects changed the modalities of the agricultural development capacity of local government officials. Whereas networks and power relations linking county and township governments with rural food processing enterprises seemed to constitute the main source of state capacity for agricultural development in the 2000s, the pull for upstream integration and DP led to the building of more intricate transversal rural *and* urban state-enterprises nexus. Today, the scheme includes not only county and township governments but higher levels as well, such as prefectural, provincial, and central officials. High-level officials sometimes express the wish to escort retailers on the field—for instance, when these latest are looking for suppliers, doing audits, or conducting trainings in rural areas— and through such visits are likely to gain political credit.

The sometimes strong involvement of central, provincial, and prefectural officials can be, as stated by an interviewee working in a retail company, "at the same time, a chance and a break." For instance, the fact that governments grant retailers with a list of suppliers in rural areas can become complicated when problems linked to food safety are encountered—whether problems are revealed by audits or discovered by a consumer, once products are on shelves. As was saying a manager of DP projects in a retail company:

> We audit suppliers [which are recommended to us by the government]. For fruits and vegetables, it is OK. For beef, it is OK. But if we find a problem for pork, it is better that we do not say we found a problem. We will sort this out by telling to the supplier that its products are too expensive, or something else. (Interview, Shanghai, October 2012)

Food safety, in China, is a very politically charged issue. The degree of political sensitivity varies depending on products. Pork, for instance, is one of the most "affected" products, given its importance in Chinese food diets—both in price and volume—and given the high risk of pork safety issues for human health.

Another issue resulting from the presence of high-level officials in the countryside is that it can complicate the mission of DP projects managers looking for suppliers. As was stating one of them:

> On this project, I have the support of the government, which is, at the same time, a chance and a break: a chance because suppliers, knowing that the government is behind the project, will be more frightened and might better fulfill their commitments [...]; a break because I am not free of doing what I want to do. [...] We had five days to visit five suppliers. Suffice to say they could tell us whatever they wanted, we had two and a half hour per supplier and they weren't going to show things that weren't working out in their companies, in front of the government. (Interview, Jiangxi, October 2013)

Still, it is essential for urban retailers to develop *guanxi* with government officials of national, prefectural, and provincial levels. *Guanxi*—especially for foreign retailers—are indeed a barrier against media attacks, which have become widespread since the 2008 melamine milk scandal. The risk that media cover food safety problems discovered in supermarkets (even if these latest are just errors in labeling or products that passed their "best by" dates but are still on the shelves) is a particularly worrying threat for them, as they face important competition and have to answer the rapidly changing demands of consumers highly concerned about food safety. The fact that food safety has become a politically charged issue and the state capacity to influence media and consumers' associations grant the government with powerful control tools over retailers. A final control mechanism highlighted by fieldwork was commercial land leases, of which the price is regularly reevaluated—every 15 years in cities I investigated. This can be problematic in overpopulated cities, where space is increasingly scarce and where renting prices can escalate dramatically.

In addition, retailers willing to integrate upstream may sometimes need the help of local governments, if not in finding local enterprises (which they sometimes prefer to look for by themselves to avoid issues discussed earlier), at least in finding local technical experts. As was saying a manager in charge of implementing DP projects in Jiangxi:

> It is important to be connected to the government. For example, we ask to the government to provide us with local technical experts. We want people who know well the area, because I will not say [to my suppliers] "do not spread this type of pesticide" and I am incapable of telling them which pesticides they have to apply, in which amounts, so we are looking for local technical experts, which are provided by the government because each local government has its own program to improve practices. (Interview, Jiangxi, October 2013)

To sum up, the eagerness of urban retailers to please the government and to get a number of resources likely to facilitate their upstream integration pushes them to establish and maintain *guanxi*. For urban governments, getting in touch with retailers is a way to have access to political credit through their involvement in DP projects.

Over the past decade, urban retailers came into the picture, as well as municipal governments having them within their area of jurisdiction. Because of the political nature of food safety issues, officials from a number of ministries (and especially from the Ministry of Commerce) also integrated the scheme. Even though the different levels of the state have the capacity to keep control on the wide variety of stakeholders of the whole food processing and retail chain, government bodies act independently from each other and defend a number of interests that they do not necessarily share with other officials. In addition, although control mechanisms exist in the hierarchy of public authorities (Chung 2010: 137), higher levels of government officials do not necessarily exert control over lower stakeholder in the food chain. Each stakeholder is in fact an individual in a wider scheme, where interests are the main push and pull factors for action (retailers being motivated by governmental incentives for DP projects, but having already started looking for direct suppliers in rural areas before the central government gave them impetus to do so). What holds the different actors of the Chinese state together? Why local officials, in rural areas, actually comply with the agricultural development guidelines promoted by the central state?

4.3   UNITING THE FRAGMENTED STATE
AROUND A COMMON MODERNIZATION FRAMEWORK

As stated earlier in the text, the decentralization of the state granted local officials with important flexibility in policy implementation. As a consequence, the details of agricultural modernization policies vary greatly from one region to another, both in institutions and in formal and informal rules governing agricultural modernization. However, interviews and policy analyses conducted in the framework of this research showed that common elements were repeatedly found in the modalities of implementation of agricultural modernization. Similarities exist both between local political discourses and between central and local discourses. These common elements progressively built a Chinese "agricultural modernization frame of reference" holding together the fragmented actors of the Chinese state. Drawing on the analysis of Number One Documents (from 2004 to 2014, 10 out of 11 Number One Documents focus on agricultural and rural development) and on fieldwork interviews (both central and local), the following paragraphs identify the main similarities between central and local discourses. The elements of what became a common framework for agricultural modernization are of two main kinds: objectives and levers.

4.3.1   A Twofold Objective

The purpose of public policies [is] no longer just to solve problems but to construct frameworks for the interpretation of the world. (Pierre Muller, *L'analyse cognitive des politiques publiques: vers une sociologie politique de l'action publique*)

The most obvious "common elements" of political discourses related to the Chinese agricultural modernization are linked to its goals. Agricultural modernization, as presented by central documents and local governments, indeed aims at fulfilling two main goals: raising agricultural production levels (especially for grain production), and increasing the income of farmers and improving the living standards of rural dwellers, partly to ensure social stability and partly to find out new levers for national economic growth.

When concerns about agriculture and rural development started re-emerging in the middle of the 2000s among officials of the central government, an important focus was put on the necessity to improve the living conditions of rural dwellers. In fact, this was the most emphasized objective in the first Number One Document linked to agriculture and rural areas. At the beginning, Number One Documents mostly insisted on the necessity to diversify income sources and to protect the legitimate rights of rural migrants (farmers working in cities or in the industrial sector). Policy guidelines then progressively started putting stronger emphasis on the necessity to protect the land rights of farmers, to improve social services and establish social security in rural areas, with the apparent intent to ensure social stability in the countryside. This goal grew stronger over the years and became a central point of Number One Documents at the beginning of the 2010s. Poverty alleviation guidelines are also included in more than half of the documents, as one of the tasks that need to be achieved to improve the living conditions of rural dwellers.

The necessity to improve the living conditions of rural dwellers and to raise farmers' income was clearly mentioned as an important objective by local officials as well. Local officials of county and township-levels extensively talked about the benefits for rural dwellers of *san nong* policies they were implementing in their area of jurisdiction. As an employee of the Investment Bureau of Lanshui told me:

> Farmers enjoy benefits from local policies. Wheat and corn production are subsidized by the government. Apples are not subsidized, but farmers do not pay taxes. Finally, farmers enjoy the benefits of a lot of policies that contribute to raise the interest of people into their products: with festivals for instance, more clients get interested into [Lanshui]'s apples and want to buy these products, thus prices rise and farmers' income rise too. (Interview, Shandong, November 2012)

The director of the Investment Bureau of Lanshui added:

> Today, our country does not have any agricultural taxes. People sell their products by themselves and every income they get from it is theirs, the government does not earn a *fen*.[3] [...] The government heavily sustains agricultural development. Each year, Number One Documents talk about agricultural issues, *sannong*. The government attaches great importance to it, and invests a lot in it every year. (Interview, Shandong, November 2012)

For local officials, raising farmers' income is essential to promote economic development and to maintain social stability. As was saying the director of the Investment Bureau of Lanshui:

Agriculture is the foundation [of China], there are many people living out of agriculture ("中国的农业人口最多 *zhongguo de nongye renkuo henduo*"), [agriculture is linked to] rural stability, national stability. If rural areas are not stable, the country will not be stable, this is why all governments have always been actively supporting agricultural work. [...] The gap between the rich and the poor is too wide, there are outstanding social problems. China is now faced to such a situation.

He added later:

The problem of China is development. [...] For economic development, we need to improve rural areas. [...] Because raising living standards, ensuring medical treatment, giving employment opportunities and improving the education of children is a necessity for the development of society. (Interview, Shandong, November 2012)

On the opposite, the willingness to increase the revenue of farmers as a way to find new levers for national economic growth was barely mentioned by local officials, who were usually much more concerned with social stability threats in their area of jurisdiction. However, according to central level officials and researchers I interviewed, rural development policies clearly aim at freeing the consumption capacity of the hundreds of millions of rural dwellers, especially given the current context of a decrease in the national economic growth rate.

Production of grain, and particularly rice and maize, has been given important attention in central documents. The necessity to develop a modern and productive agriculture gradually grew stronger in the documents, before reaching a peak in 2014, when the Number One Document devoted an entire chapter on the "necessity to improve national grain security protection system", given the "new circumstances"—basically, the rapid increase in grain imports since 2004 and the growing deficit of the agricultural trade balance. Over the years, emphasis was also gradually put on the necessity to increase the production of other agricultural commodities. The 2005 Number One Document, for instance, wishes to develop animal husbandry. In 2007, the focus is then put onto aquaculture, before political guidelines, starting from 2008, start including recommendations to increase production levels of commodities of a diversified food basket, including vegetables, meat, and fish (Table 4.3).

The necessity to develop a modern and productive agriculture was a wish expressed by local officials as well. In Lanshui, a lot of policies and

Table 4.3 Emphasis put by 2004–2014 Number One Documents on grain and other agricultural commodities production (occurrences in paragraph titles and subtitles)

	Support and increase grain production levels	Support and increase other agricultural commodities production levels
2004	1. "Support main grain producing areas and grain industries' development and increase grain-growing farmers' income" 7.b) "Deepen the reform of the grain distribution system"	*Not mentioned neither titles nor in subtitles*
2005	1.b) "Reinforce support for the major grain producing areas" 6.a) "Go a step further in improving grain production" 6.d) "Sustain the development of processing capacities in major grain producing areas"	6.c) "Accelerate the development of **animal husbandry**"
2006	2.c) "Stabilize grain production"	*Not mentioned neither titles nor in subtitles*
2007	4.a) "Promote the stable development of grain production capacities"	4.b) "Develop healthy **aquaculture**"
2008	2.a) "Attach importance to the development of grain production"	2.b) "Improve the production of the **whole food basket** (including vegetables, meat and fish)"
2009	2.a) "Vigorously support grain production"	2. b) "Sustain **oil and cash crop** production" 2.c) "Accelerate the development and standardization of **animal husbandry and fishery**"
2010	2.a) "Steadily develop the production of grain and other staple products"	2.b) "Push forward the standardization of production of **vegetables** and other products"
2012	1.a) "Keep up efforts for grain production"	1.b) "Pay close attention to production of **vegetables** and other products"
2013	1.a) "Steadily develop agricultural production"	
2014	1. "Improve national grain security protection system"	*Not mentioned neither titles nor in subtitles*

programs were implemented to develop local food production. Subsidies for grain production were mentioned by the interviewees, as well as other policies targeting apples—the main agricultural output of the area—and other products. In Lanshui, the county government had decided to establish a fruit development bureau, as a way to provide technical answers addressing the issues encountered by local farmers. In Lushan, the local government was pushing enterprises to train farmers so that they could increase their yield. I could see many pest traps in the orchards and was told that they were heavily subsidized by the government. The active promotion of local fruits throughout county governmental bureaus, both in Lushan and in Lanshui, was also part of the strategy of local governments aimed at helping the development and modernization of the agricultural sector—even though fruits are far from being the first priority set by central documents.

4.3.2   *Favored Levers: Technology, Industrial Actors, and Rural Exodus*

Frames of reference for modernization are not only defined by the goals that modernization policies put emphasis on. Frames of reference promote levers as well, to reach the objectives that they support. In the case of agricultural modernization, three main levers were regularly promoted to modernize the sector.

The first lever is science and technology. Science and technology are really at the core of Chinese central discourses on "modern agriculture" and clearly appear at local levels as well. It is interesting to note that this faith in science and technology, which is regularly expressed by officials in documents and in interviews, strongly echoes the faith that the society has in science and technology. The results of the World Value Survey 2014, for instance, show that to the question "Science and technology are making our lives healthier, easier, and more comfortable," 73 percent of Chinese respondents said that they strongly agreed.[4] In the agricultural sector, scientific and technological modernization includes a wide range of techniques, from the most basic ones (e.g., agricultural machinery, pesticides, and fertilizers) to the most elaborated and capital-intensive ones (e.g., improved seeds). The fundamental role of science and technology for agricultural development is mentioned in all Number One Documents since 2004. Among other things, strong emphasis is put on the development of research capacities. Researchers met during fieldwork confirmed that considerable financial efforts had been put in the development of research centers, which are today equipped with cutting-edge technology. As far as rural areas are concerned, mechanization and informatization are regularly mentioned in central documents as well as in local governments' discourses. As was stating the director of the Investment Bureau of Lanshui:

> The government is attaching strong importance to agricultural mechanization. […] If you want to purchase modern agricultural machinery, the government will give you subsidies, in order to encourage you to use advanced technology and equipment. In the past, Chairman Mao used to say that the basic foundation of agriculture was mechanization. He was already aware of this issue at that time. (Interview, Shandong, November 2012)

While wandering in the countryside, it happened a lot that people who knew that I was working on agricultural modernization showed me tractors and said "See! Agricultural modernization!"

However, technological progress has today proven to be useless without technological extension services. The inefficiency of the overuse of chemical fertilizers and the damages it has on the environment clearly illustrates the issue. Technological extension is mentioned in Number One Documents, but with less emphasis compared to the one put on the development of research capacities and technology industries. During the first half of the 2000s, considerable efforts were dedicated to the development of upstream research facilities and industrial capacities, while few concentrated on how to link the final users of technology—farmers.

In the second half of the 2000s, however, the need to "foster rural talents" and to "breed a new variety of farmers" gradually emerged. According to an employee of the government of Lanshui:

> The government developed training programs for farmers. There is a fruit development bureau in the government. They have more than ten senior agricultural experts who teach at a fruit tree station. They also put a lot of efforts into the upgrade of technology in industry. (Interview, Shandong, November 2012)

However, the incapacity of local government extension services to answer the specific needs of farmers was regularly denounced by farmers and by industrial players and NGOs closely working with them. It appears that a lot of progress can still be achieved to improve how technology manages to reach farmers.

If the role that "talented farmers" can play in agricultural modernization is recognized by some of the 2004–2014 Number One Documents, the role industrial stakeholders are encouraged to take on is much more strongly and more frequently emphasized by the same documents. This constitutes the second favored lever for agricultural development. The 2004 Number One Document, for instance, pushes "dragonhead enterprises"[5] to "provide trainings and marketing services to farmers, to feed agriculture with new technology" and to take on a number of similar "leading" roles. In the 2007 Number One Document, "dragonhead enterprises" are mentioned as key players "leading farmers' development and agricultural modernization." The importance granted to enterprises was found at the local level as well. According to an employee of the Investment Bureau of Lanshui, the living conditions of farmers had "improved a lot here, thanks to the food enterprises who invest and buy their products." (Interview, Shandong, November 2012) Fieldwork conducted in Jiangxi, as well, demonstrated that the willingness of local officials to grant industrial players—and not only dragonhead enterprises—with a leading role in agricultural modernization was extremely strong—for a number of reasons depicted in Chap. 3.

On the opposite, grassroots organizations play a relatively small role in agricultural modernization. Even though the 2006 Number One Document writes about the necessity to breed "new types of service organizations"—other than the collective ones—at that time, only professional associations were mentioned, of which the members are usually food industries and traders. In 2007, the necessity to promote the development of agricultural cooperatives emerged as a new lever to provide services to farmers. In 2012, the role of rural associations expanded to cover a wider variety of services, from financial services to marketing or technology extension. All forms of rural organizations, from then on, were encouraged: agricultural cooperatives, supply and marketing cooperatives, technology associations, water associations, enterprises, and so on. In 2013 and 2014, the necessity to develop all forms of rural organizations is again emphasized in central documents.

As we see, from 2004 to 2014, the exclusive leadership of enterprises in agricultural modernization gradually evolved and started integrating other stakeholders as well as "grassroots" forms of organizations. However, as documents underline it, rural cooperatives only complement—and never replace—dragonhead enterprises as service providers to farmers. In the areas where I conducted fieldwork, enterprises have remained the most important players, in spite of the recent change in central policy guidelines. This topic, which deserves to be discussed more thoroughly, will be further investigated in the following chapter.

The last element defining the Chinese frame of reference for agricultural modernization is the idea that labor migration out of the farming sector and rural exodus are important *levers* for the increase in agricultural productivity and for the improvement of the living conditions of rural dwellers. For instance, the 2004 Number One Document argues that pushing more farmers to live in small towns will have positive effects on industrial development, population gathering, and market enlargement. The 2006 Number One Document expresses the willingness of the central government to establish rural–urban networks of public services able to provide free information, guidance, and assistance to former farmers willing to work in the industrial sector. Before 2008, central documents used to put strong emphasis on the necessity to protect the legitimate rights of migrant farmers (i.e., the ones taking jobs *outside* the farming sector). All these guidelines clearly intend to facilitate rural exodus.

In 2008, along with the goal of protecting migrant workers, another goal emerged: the one according to which the rights of farmers staying in rural areas should be protected as well. In 2008, only farmers' land rights are clearly mentioned in the titles of subparagraphs. In the following years, additional features such as forest collective rights or land contract reform were added to policy guidelines. However, the rising necessity to protect farmers' rights did not lower the eagerness of the state to encourage rural exodus and migration of labor out of the farming sector. The 2013 Number One Document clearly stipulates that the "urbanization of farmers" should be encouraged (in particular, through the relaxation of the *hukou* systems of small and medium towns, the establishment of social security and assistance for migrants, etc.). Similarly, the 2014 Number One Document states that the urbanization of farmers should be accelerated. Five-Year Plans, on their side, keep on promulgating urbanization rate targets.

Such a discourse was clearly found at local levels as well. When I was noticing the old age of the agricultural labor force (especially in Ningxia and, to a lesser extent, in Jiangxi and Shandong), I systematically asked questions about local rural–urban migration policies. Answers were invariably defending the same logic: rural exodus is a positive process, because farmers staying in rural areas will be able to have bigger farms and earn more money. According to the director of the Investment Bureau of Lanshui:

> The land per capita is very small, three mus or two mus, four-five mus is already a lot. As a consequence, it is very difficult to manage the shape of farm. If we want to change the mode of agricultural production in the future, we have to concentrate the landholdings, in order to have owners of big farms. When leasing markets will be established, landless peasants will take temporary jobs, it will be modern farmers, it will more convenient to manage and there will be technological upgrading. This is the path for the future. (Interview, Shandong, November 2012)

In the county of Huangmo, in Ningxia province, agricultural investors were almost nonexistent at the time when fieldwork was conducted. Other agricultural development models emerged and spread across the county, where a wide variety of stakeholders are involved, from county- and township-level officials to village leaders, government associations, NGOs, and enterprises (Table 4.4).

However, even though the lever of industrial players could not be part of the discourse of local officials in Huangmo, these latest were keeping on referring to the other elements of the dominant frame of reference: the lever of technology and the lever of rural exodus. For local officials, modernization mainly refers to an increased use of technology, as this quote from the secretary (书记 *shuji*) of a township in Huangmo illustrates:

> In France, you have cellphones, right? Well, this is modernization. (Interview, Huangmo, June 2013)

In addition, urbanization is considered as a key lever for agricultural development, even though local conditions do not lend itself to it, as the majority of people staying in rural areas are above 50 or 60 years old. According to the same township secretary:

> There are national urbanization targets. It is not problem if a share of the farming population goes to the city. People who stay here are encouraged to do family farming (家庭农场 *jiating nongchang*). (Interview, Huangmo, June 2013)

Table 4.4 The three models of agricultural development in Huangmo county

Governmental Associations and NGOs

An interesting actor involved in rural development is the county "association" for rural sustainable development ([Huangmo]县农村可持续发展协会 [*Huangmo/xian nongcun kechixu fazhan xiehui*]). Registered in 2006 with the Bureau of Civil Affairs, the structure in fact already existed before and used to run projects under another name ([Huangmo]县农业产业化网络协会 [*Huangmo/xian nongye chanyehua wangluo xiehui*, [Huangmo] association for rural industrialization network). Although the translation for *xiehui* is "association", the *xiehui* for rural sustainable development of Huangmo is in fact closer to a governmental structure than to the one of an association. Firstly, the bureau of the association is located inside the official buildings of the Huangmo government. In addition, at the time when interviews were conducted, five people were working at the *xiehui*, among whom three—including the secretary general—were government employees (政府官员*zhengfu guanyuan*) (Interview with the secretary general of the Huangmo, June 2013).

The *xiehui* operates on a membership basis. At the time when I conducted interviews, the *xiehui* had 60 individuals (farmers农民*nongmin*) and 26 nongovernmental organizations (非政府组织 *feizhengfu zuzhi*) registered as members. In appearance, the structure of the *xiehui* is thus similar to the membership structure of grassroots associations. However, the mode of operation of the *xiehui* differs significantly. Half the managers of the association belong to the government and are responsible for "selecting" villages for development projects. The mode of operation of the *xiehui* is thus far from the one of grassroots associations or nongovernmental associations and seems closer to a traditional top-down management scheme (quite similar results were found by political scientists having conducting research in other sectors of the economy. Kenneth Foster (2002: 43), for instance, argues that associations are "new elements of the state's administrative system").

The association supports two types of projects, usually proposed by its "NGO members": (i) short-term trainings for agricultural technology extension (农业技术培训 *nongye jishu peixun*); (ii) small interest loans projects (低息贷款 *dixi daikuan*), among which many support farmers. Most of the NGOs conducting projects in Huangmo come from external areas and lack local contacts and information on the context. They come seek the help of the *xiehui* for information and contacts they would otherwise have strong difficulties to gather by themselves, because of the remoteness of villages and because of local circulation constraints. The *xiehui*, on its side, closely related to the government, has a wide network of governmental contacts and can easily link NGOs with local government officials—without whom nothing could be done. The control over the uncertainty of local conditions and local networks grants the permanent staff of the *xiehui* with significant power over its NGO members.

County-level officials play a leading role in the daily activities of the *xiehui*. In addition, the creation of the *xiehui* itself seems to be stemming from policy guidelines. In 2001, the central government published a white paper on rural poverty alleviation and development. The document emphasized the necessity to "actively create the conditions to encourage non-governmental organizations to participate in and carry out governmental poverty alleviation projects" (State Council Outline for China Rural Poverty Alleviation and Development, 2014). The county of Huangmo, as a former revolutionary base suffering from difficult economic conditions, was particularly encouraged to follow these guidelines, which would explain the creation of the association for rural industrialization network, and, later on, its evolution into the association for rural sustainable development.

(*continued*)

Table 4.4 (continued)

Officials-Entrepreneurs

Another interesting model of agricultural development, in Huangmo, was the one of agricultural cooperatives. Agricultural cooperatives are usually strongly invested by private stakeholders in China. In the county of Huangmo, in the absence of investors, agricultural cooperatives I investigated were created and headed by township-level government officials, such as this leader of a farmers' cooperative, who was explaining:

I decided to create this cooperative some years ago, in order to help the common people (老百姓*laobaixing*). I provide them services such as joint purchasing of fertilizers, so they can have access to cheaper products. I help them sell their products. Here, peasants do not use pesticides, so we promote a green brand for potatoes we sell. [...] I also provide them with training and education. [...] The cooperative developed and reached a good production level for potatoes. (Interview, Huangmo county, June 2013)

Such an embeddedness of local political leaders within the Chinese economic sector was described by a number of scholars, such as the ones defending the theory of entrepreneurial state or the theory of developmental state. Jane Duckett (1996: 3), for instance, argues that in the 1990s, "individual departments across the state system have been setting-up new profit-seeking, risk-taking businesses". Marc Blecher and Vivienne Shue (1996) also found areas where administrative agencies of county governments were conducting development activities in order to achieve economic profits for themselves. However, most of the research on the profit-seeking action of local officials in economic development focuses on the development of the industrial sector in the 1980s and 1990s (Byrd and Gelb 1990; Oi 1992; Walder 1995). Much less scholars concentrated on the involvement of local officials in agricultural activities, and many described this involvement as a direct descendent of the collective management system (Chen 1998; Wong 2016), not as a self-breeding new form of entrepreneurship. The involvement of local officials in agricultural cooperatives in Huangmo proves that the model depicted by the theory of entrepreneurial state is still valid in some rural areas and can be found in the agricultural sector as well, independently from the past of rural areas in terms of collective agriculture.

In Huangmo, enterprises are far from being the major players in agricultural development and play a little role in the development of agricultural cooperatives. As was stated by the secretary general of the rural sustainable development association:

Among our members, there isn't any enterprise. Some enterprises belong to *xiehui* in China, but in our *xiehui*, there isn't any, because they don't want to invest money in agriculture in [Huangmo]. (Interview, Huangmo county, June 2013)

(continued)

Table 4.4 (continued)

Microcredit Enterprises

The objective of H. enterprise is to offer microcredit for farmers in the county of Huangmo. Although the enterprise has strong links with the *xiehui* mentioned earlier in the text, employees manage to operate relatively independently from county government officials. The usual procedure is the following one: employees of H. first go to villages through their own means and seek the support of the village leader (总主任 *zhongzhuren*). Employees admitted that village leaders were governmental employees, but insisted on the fact that they were social actors first, because most of them were actually farmers. Village leaders are then supposed to disseminate information to isolated rural dwellers in the village and try to win their support to launch microcredit. In the second phase of the process, H. proposes to local relays, who have played a role in the development of local networks in villages, to manage these latest on a permanent basis, on the behalf of the enterprise. The possibility for local people to access to a position with responsibility in the enterprise enables the building of stable relays and facilitates the establishment and maintenance of consolidated local networks. Thanks to this process, H., at the time fieldwork was conducted, employed 16 local relays (信贷员 *xindaiyuan*, "credit personnel"), through whom the enterprise could provide microloans to about 3000 farmers.

The status of H. is ambiguous. H. used to be a *xiehui* (妇女发展协会 *funü fazhan xiehui*), then turned into a "nongovernmental *danwei*" (status close to the one of NGO) (扶贫与环境改造中心 *fupin yu huanjing gaizao zhongxin*) in 2004, before becoming an enterprise in 2008. The evolution of H. into a private enterprise was motivated by a need to get more funds to become able to provide loans to more clients. Even though this evolution is still lived and described as a difficult step by the managers of the enterprise, the seeking of financial partners is in fact a sign of development, which demonstrates the success of H. and of its usefulness for farmers in rural Huangmo, deserted by investors. At the time when I conducted interviews, H. was one of the sole microcredit enterprises, throughout the whole country, that provided credit to farmers with the aim of helping them develop agricultural activities. Most of microcredit enterprises, in China, indeed still focus on the development of commercial and industrial activities for poverty alleviation in rural areas. However, the experience of H. proves that this model could be an interesting lever for development.

Or, according to the leader of the farmers' cooperative quoted in Table 4.4:

> People living for cities are not posing a problem, because farmers cultivate very small areas. People who stay can take care of these areas. (Interview, Huangmo, June 2013)

As these quotes illustrate, even though agricultural development models are different in the county of Huangmo, the ideas local officials share about modernization are quite similar to the frames of reference acknowledged in other areas. Strong similarities thus exist between central level documents and local discourses, both in terms of goals and levers. How come the elements of the central frame of reference of agricultural modernization are transmitted down to local levels of public authorities? The following subsection aims at providing some answers to this question.

4.3.3   Spreading the Framework

A large body of literature focuses on the question of transmission mechanisms in the fragmented Chinese state. According to Chien (2010: 137), several mechanisms are used by the central government to control the lower levels of public authorities: (i) administrative orders; (ii) mandatory plans (plans for social and economic development, measured by indexes such as GDP and FDI); (iii) allocation of financial and other resources; and (iv) personnel appointments and removals. Smith (2009: 30), on his side, argues that county governments have important leeway in policy implementation and "only take up initiatives wholeheartedly when three conditions are met: (i) the initiative is important to the annual assessment system; (ii) the initiative raises revenue, either through levying fines, taxes or service fees, or by opening up revenue sources from higher levels; (iii) the initiative benefits individual cadres and the 'shadow state' financially". These valuable explanatory frameworks, however, did not completely correspond to what I could observe in Jiangxi and Shandong. In these areas, strong similarities could be found between central and local frames of reference for agricultural modernization, even though the implementation of central policy guidelines was neither generating additional revenue nor enabling local cadres to be better ranked in the cadre evaluation system. This does not mean that the cadre evaluation system does not play any role in the transmission of central agricultural modernization guidelines. However, the fieldwork of this research proved that other mechanisms, both direct and indirect, were important as well in the whole system allowing the transmission of the frame of reference down to local levels.

In the literature, the Chinese cadre evaluation system was depicted as an important steering mechanism for upper-level public authorities (Heberer and Trappel 2013; Li and Zhou 2005; Edin 2003; Gao 2009). Under the system, which is implemented at each level of the government, officials are evaluated by the Organization Department and the Party Committee of the level just above their own. Targets, which set by evaluating offices, are ranged on a grading scale according to their relative importance: soft targets (软指标*ruan zhibiao*), for low-priority tasks; hard targets (硬指标*ying zhibiao*), more important to achieve; and "one vote down" targets (一票否决*yipiao foujue*), of which the failure automatically results in punishment and cannot be remedied by good achievements in other areas. Some targets are quantifiable and evaluated through measurable figures (e.g., GDP and birth rate), whereas others cannot be assessed through specific indicators ("integrity", "incorruptness", etc.).

Family planning, social stability, and economic development are traditionally considered as critical tasks that have the largest impact on the career of officials (Burns and Zhou 2010; Li and Zhou 2005), even though in recent years, economic targets were the subject of intense debates (People's Journal 2013). The strong emphasis that has always been put on social stability and economic growth in the cadres evaluation system facilitates the transmission of the objective of "improving living conditions in rural areas". Almost all Number One Documents, since 2004, have stressed the fact that improving rural infrastructures was necessary to create good conditions for economic development. In addition, Number One Documents from 2004 to 2007 emphasize the need to promote economic growth in small towns and to diversify the income sources of rural dwellers. From 2008 on, as concerns about rural social stability issues linked to farmers' expropriation grew stronger, Number One Documents started focusing on the necessity to protect farmers' rights, to reform land property system and to fill the gap between the living conditions in rural and urban areas. As we can see, economic growth targets—and, later on, social stability targets, because they are critical for the evaluation of cadres—encouraged local officials to carry out policies complying with the rural development guidelines promoted by the central government.

However, the traditional cadre evaluation system is far from being sufficient to explain the eagerness of local officials to implement agricultural modernization policies. Activities other than agricultural production can indeed contribute to social stability and economic growth in rural areas way more than agricultural development does—industrialization, for instance, has long been the preferred way economic development in rural areas. The traditional evaluation system is in fact complemented by other evaluation mechanisms. Local agricultural and grain bureaus, for instance, are responsible for checking the enforcement of grain production targets. The fundamental importance attached to national grain self-sufficiency led to the establishment of the "governor's grain bag responsibility system" in 1995. Provincial governors are in charge of balancing local grain supply and demand, by supporting grain production. In addition to the governor's grain bag responsibility system, grain production targets are set every five years by the central government in Five-Year Plans. These targets are then progressively detailed by each level of the government, down to local grain bureaus, which set local grain production targets on a yearly basis.

Five-Year Plans also include production targets for other agricultural products, such as oilseeds, sugar, meat, and dairy products (Table 4.5). Although reaching these targets is less important than reaching grain targets at the national level, it can matter in some provinces that specialize in these kinds of products. In addition, another system encourages government officials to take vegetable production seriously: the mayors' vegetable basket, which was implemented by the MOA in 1988. Under this program, local agricultural bureaus have built thousands of wholesale agricultural markets in order to improve the production and marketing of vegetables and other food products. It is still and important system today that encourages areas to consume local vegetables.

Table 4.5 12th Five-Year Plan agricultural production targets

Target	2010	2015	Annual rate of increase (%)
Grain overall production capacity (100 million tons)	>5.0	>5.4	
Grain sown area (100 million mus)	16.48	>16.0	
Cotton total output (10,000 tons)	596	>700	>3.27
Oil seeds total output (10,000 tons)	3230	3500	1.62
Sugar products total output (10,000 tons)	12,008	>14,000	>3.12
Meat total output (10,000 tons)	7926	8500	1.41
Egg total output (10,000 tons)	2763	2900	0.97
Dairy products total output (10,000 tons)	3748	5000	5.93
Fishery total output (10,000 tons)	5373	>6000	>2.23

To sum up, the combination of responsibility and evaluation systems complements the traditional cadres evaluation system. The whole scheme establishes formal mechanisms that are supposed to push local officials to keep food production at the core of the rural policies they implement in their area of jurisdiction.

The existence of this set of agricultural production targets, however, is not sufficient to explain why local governments pursue agricultural development goals. Agricultural targets are indeed not among the "one vote down" targets and can be compensated by other achievements, which can at the same time grant local officials with greater financial and political power. As are noticing Zha and Zhang (2013: 462):

> Although the central government is committed to ensuring grain security for the nation and promoting farmers' incomes, the local governments show little interest in the agricultural sector [...] Agriculture does not help the local government's promotion system. Promotion of local government officials is strongly based on merit, especially their contribution to economic growth. However, agriculture, particularly the grain sector, generates little employment for the local economy and its contribution to GDP growth is negligible. (Zha and Zhang 2013: 462)

Göbel (2011: 54), on his side, observes that the hardness of targets cannot entirely explain the efficiency of policy transmission, because despite the fact that "local leaders everywhere face the same targets", "one of the same policy often produces eager supporters (known as 'pioneers') in one locality and resisters in another". Finally, agricultural targets already existed before the 2000s, whereas conclusions drawn from fieldwork show that local governments only started renewing their interest in agriculture and rural areas about a decade ago. Therefore, other explanations need to be found.

The competitive environment within which local officials evolve partly explains this puzzling issue. In a competitive political environment, agricultural development indeed becomes a strategy for "marketing differentiation" for local officials. Landry (2008: 21) states that "local competition [breeds] political competition by creating local power bases that undermine political cohesion." It is true that on my fieldwork, I could see that political cohesion was not the best strength of local governments. However, this lack of cohesion between the bureaus of a same administrative level was not impeding the implementation of agricultural development policies. On the opposite, the competitive environment between local officials encourages these latest to adhere to central policies, because their results are likely to allow them to have access to political credit in front of higher level officials, potentially leading them to higher positions in the hierarchy of the administration. The willingness of municipal officials to be involved in DP projects and to go to rural areas illustrates this point. The agricultural development project conducted by the poverty alleviation bureau (without involving the agricultural bureau) in the county I explored near Chongqing is another example of the efficiency of such a competitive environment in leading to the implementation of local agricultural development policies.

The explanation given by Göbel shares similarities with the conclusions I drew from my fieldwork. According to Göbel, the uneven implementation of rural policies is due to what he calls "competition under hierarchy", a system under which "pioneers are motivated to go along, not only by fear of punishment, but also by the promise of material and immaterial rewards" and where "resistance is the result of a locality's inability or unwillingness to engage in competition". However, fieldwork also demonstrated that this mechanism of evaluation and competition was completed by other types of control mechanisms for the implementation of agricultural development policies, among which the allocation of financial resources (Fig. 4.4). Local governments indeed highly depend on the

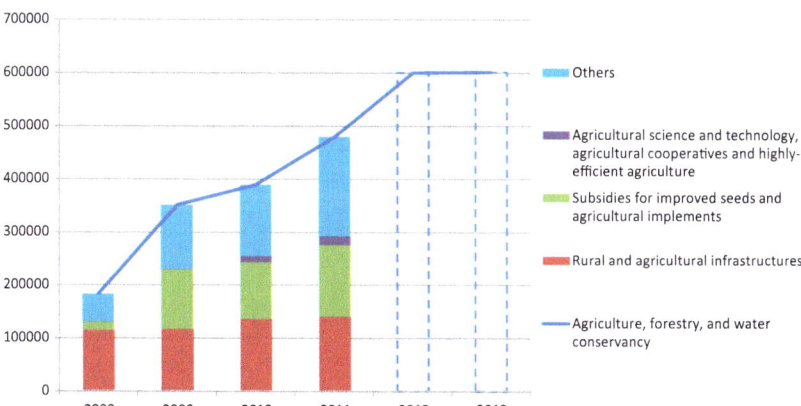

Fig. 4.4   China's agriculture, forestry and water conservancy expenditures from 2008 to 2013 (Unit: million RMB) (Source: 关于2008–2013年中央和地方预算执行情况与2009–2014年中央和地方预算草案的报告 guanyu 2008–2013 nian zhongyang he difang yusuan zhixing hang qingkuang yu 2009–2014 nian zhongyang he difang yusuan cao'an de baogao [Report on 2008–2013 central and local budget situation and 2009–2014 draft for central and local budget])

redistribution of the revenue collected by the central state (Figs. 4.5 and 4.6). According to Shen et al. (2012: 17), transfers from the central government to provinces account for 67 percent of provincial fiscal resources and transfers from provinces to subprovincial governments account for more than half of these latest fiscal resources. At the national level, agriculture is budgeted in two main items: "Agriculture, forestry and water conservancy" and "Grain and edible oil reserves and other related measures". The amount of expenditures dedicated to both items kept on increasing over the past few years. "Agriculture, forestry and water conservancy" jumped from 182,174 million RMB in 2008 to 600,540 million RMB in 2013, whereas resources allocated to "Grain and edible oil reserves" went from 46,169 to 126,638 million RMB over the same period of time[6] (see Fig. 4.6).

The two most important items of expenditures allocated to agriculture are "Rural and agricultural infrastructures" and "Improved seeds and other agricultural implements" (Fig. 4.4). According to an interview conducted with an official working on rural expenditures at the Ministry of Finance, most of the resources allocated to the improvement of infrastructures come from local governments. On the opposite, almost all of the subsidies directly allocated to agriculture come from the central level, because agriculture does not generate local revenue since agricultural taxes were abolished in 2006. Such a scheme of expenditure allocation grants higher levels of the government with an important steering mechanism to push local officials to make efforts to develop agriculture.

The steering mechanism becomes particularly powerful in townships and villages, as these latest have scarce resources and highly depend on higher levels for their revenue (Oi et al. 2012). This lack of financial capacity at the township and village levels was widely denounced as a negative consequence of fiscal reforms. Shen et al. (2012), for instance, point at the inconsistencies to which the reforms of the fiscal system led. For the authors, "the higher tiers of government devolve fiscal responsibilities down to the lowest levels of government and meanwhile the most productive sources of revenue are captured by the top tiers of government". Smith (2009: 601), on his side, notes that township governments find themselves squeezed both from above and from below.

In the areas I investigated, issues linked to the lack of financial capacity, especially at the township level, were raised as well by a number of interviewees. For instance, I was explained by a manager of a foundation conducting land planning projects in rural areas in Jiangxi that local officials

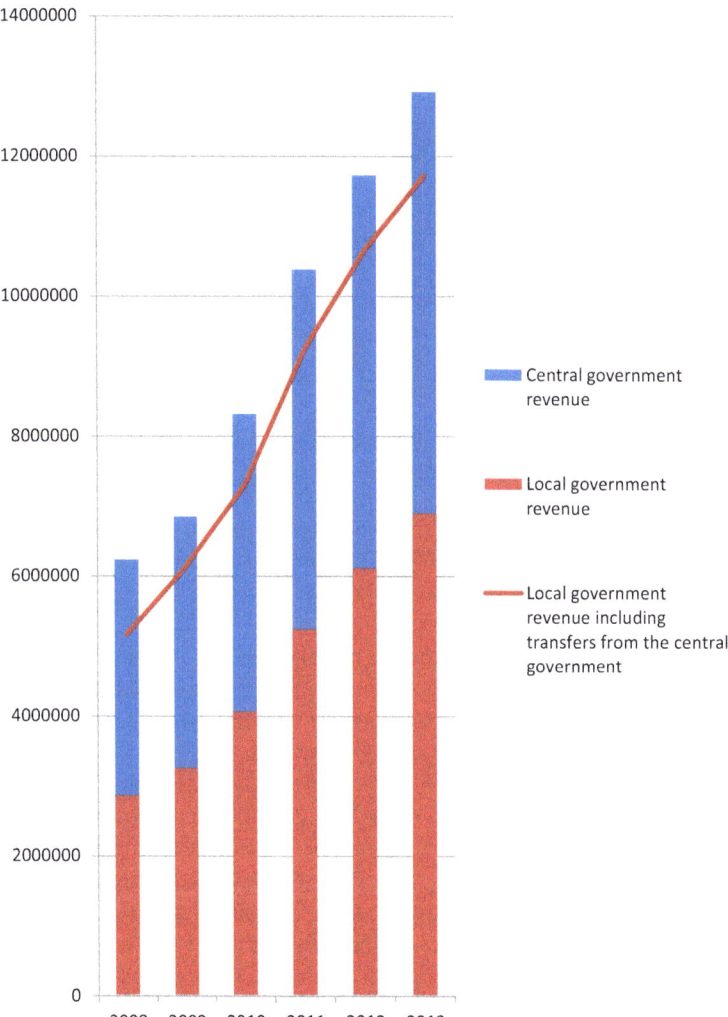

Fig. 4.5 Transfers from the central government for local government revenue (Unit: million RMB) (Source: 关于2008–2013年中央和地方预算执行情况与 2009–2014年中央和地方预算草案的报告 guanyu 2008–2013 nian zhongyang he difang yusuan zhixing hang qingkuang yu 2009–2014 nian zhongyang he difang yusuan cao'an de baogao [Report on 2008–2013 central and local budget situation and 2009–2014 draft for central and local budget])

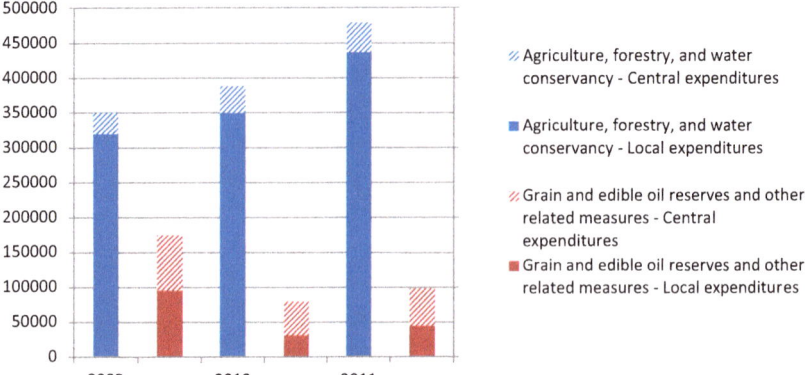

Fig. 4.6 Central expenditures for rural and agricultural development (Unit: million RMB) (Source: 关于2008–2013年中央和地方预算执行情况与2009–2014年中央和地方预算草案的报告 guanyu 2008–2013 nian zhongyang he difang yusuan zhixing hang qingkuang yu 2009–2014 nian zhongyang he difang yusuan cao'an de baogao [Report on 2008–2013 central and local budget situation and 2009–2014 draft for central and local budget])

were paid only 2000 RMB per month and were much eager to dedicate time to activities generating money (either to the production of oranges when they farmed themselves or to other activities such as trade) than to public management. However, the fact that local officials from township and county levels are not fiscally autonomous is in fact part of the steering system allowing for a transmission of the goals of the central government down to local officials. Local officials who wish to keep the same budget from one year to another need to report their expense to higher-level officials. In particular, during fieldwork, I could acknowledge the eagerness of a number of local bureaus in charge of developing the agricultural sector to spend the funding that had been allocated to them the previous year, in order to maintain their level of public funding for the following year. This was an additional incentive to encourage them to implement agricultural development programs.

4.4   Conclusion

The new role granted to private enterprises in the course of agricultural modernization does not mean that government officials were not able to establish control mechanism on the emerging forms of agricultural–industrial capitalism. In Jiangxi and Shandong, in particular, local government officials were not directly involved in the process but still played a role through their integration into state-enterprises networks, over which they established control by using existing resources and regulations. The fact that decentralization granted local authorities with considerable leeway in their use of resources and regulations led to the birth of "fragmented pseudo-regulatory local states," where officials apply rules according to their own and specific terms, in order to push and pull entrepreneurs to take part in the modernization of the agricultural production sector.

In *Capitalism from Below*, Nee and Opper (2012: 150) argue that "with the continuing expansion of markets, the economic success of firms became increasingly independent of the direct involvement of politicians". For the authors indeed, "vertical ties linking economic actors in firms with the state decline in significance as horizontal ties—interfirm networks and network ties between buyers and sellers based on repeat exchange—gain in importance". However, insights from fieldwork showed that the links with the government had neither faded nor decreased in importance. On the opposite, most of the interviewees said that even after the crucial step of land attribution, entrepreneurs remained eager to maintain strong links with government officials, in case they would be willing to expand their activity and even for the smoothness of day-to-day business. Urban retailers, as well, had to face the continuous pressure of "local governments"— as termed by them. As a consequence, the emergence of transversal networks of agrarian capitalism, involving both downstream and upstream private entrepreneurs, does not put back into question the strong capacity of the state, of which officials remain strongly integrated in the scheme of agricultural modernization (Fig. 4.7).

Two words were continuously coming back: *lingdao* ("officials") and *guanxi* ("relationships"). Contrary to a few scholars who noticed in the 1990s that the importance of *guanxi* was declining in the economy (Guthrie 1998), the fieldwork of this research shows that *guanxi* with the government (or "political capital" in the broader sense of the term) are still fundamental to food processing enterprises for their launch, survival, and economic success. Maintaining good relationships with local officials is essential because this latest grant food enterprises with resources—otherwise scarce, scattered, or nonexistent—during the implementation stage (for resources, such as land, human resources, and licenses) and thereafter (subsidies, renewal of licenses, granting of certificates, etc.). The power of local governments is increased by the fact that they can take some of these resources away from enterprises (licenses) even once they are granted.

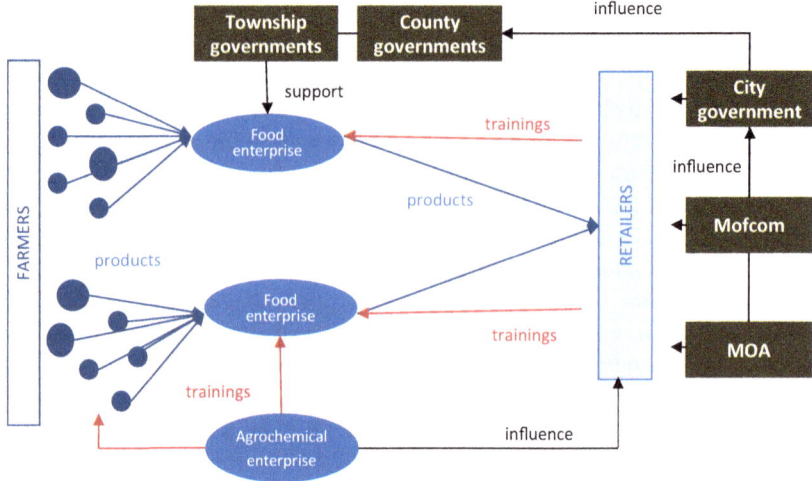

Fig. 4.7   Interventions of governments and retailers in the food chain

Even for uncertainties they control, enterprises need local governments, which can either strengthen or threaten their control over uncertainties. As was summing up a manager of a rural food enterprise:

> *Guangxi*, in China are very important. Without relationships with the government you cannot do anything. (Interview, Jiangxi, October 2012)

The aim of this research is not to argue that "the Chinese state" has reintegrated agricultural production activities. As the previous paragraphs demonstrated, "the" Chinese state is highly fragmented as well as the other actors taking part in the modernization of agriculture. State-enterprises networks of the agricultural and food sector involve a wide variety of players, among whom government officials act independently from each other and defend interests they do not necessarily share with their colleagues. This does not mean, however, that the Chinese government is a completely incoherent body. Two main goals—agricultural productivity and rural development—and three main levers to achieve these goals—science and technology, industrial players, and rural exodus—are regularly promoted by central level documents. They constitute the frame of reference of agricultural modernization as promoted by the central state. This frame of reference is found at the local level as well, in the discourse of local officials.

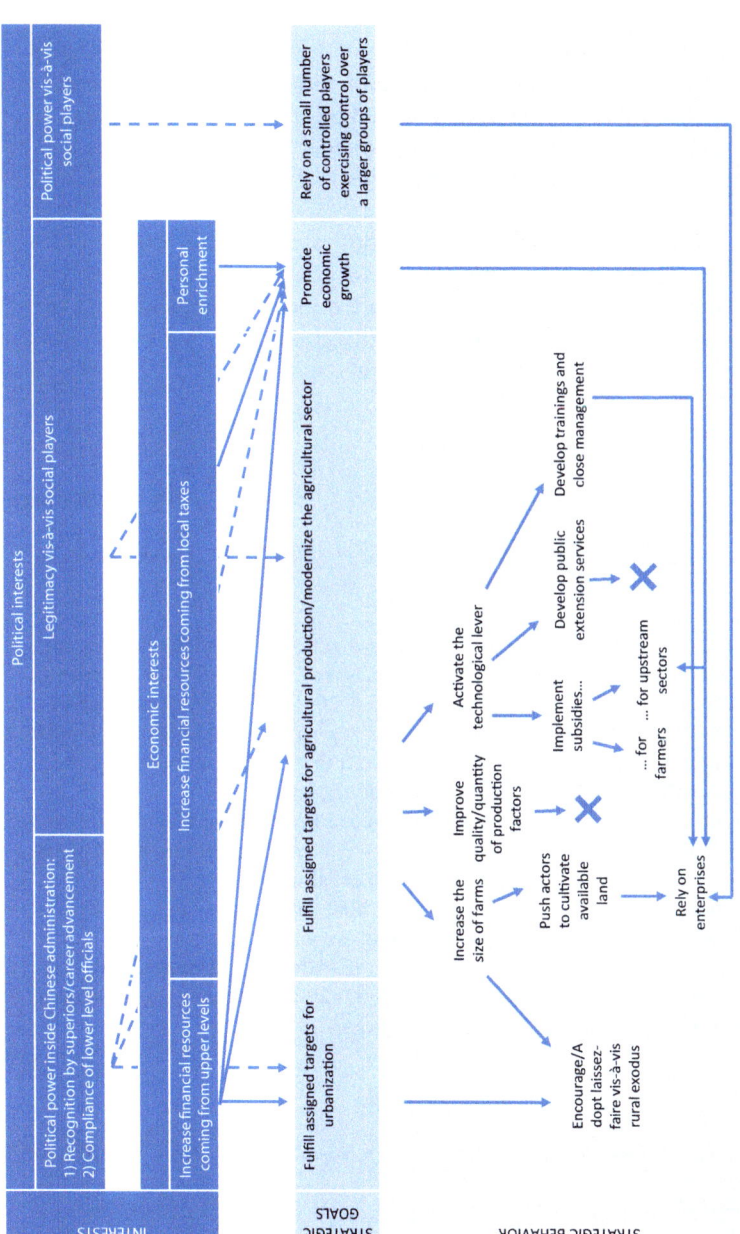

Fig. 4.8 Set of interests, strategic goals, and strategic behaviors of local governments

The transmission of the elements of the dominant frame of reference for agricultural modernization is permitted through direct mechanisms such as cadres evaluation systems and budget allocation but mostly, as fieldwork demonstrated, because these elements fit in path dependencies as well as in the current pattern of interests of local economically and politically powerful stakeholders (Fig. 4.8).

As the two following chapters will demonstrate, the most recent guidelines on democratic management and grassroots organizations, which fit less in local patterns of power, are way more difficult to implement at the local level. As we will see, this has tremendous consequences for the trajectory of agricultural modernization in China.

NOTES

1. Original language: 希望各农机企业要宣传好 (*xiwang ge nongji qiye yao xuanchuan hao*).
2. Original language: "我们是需要政府支持的,这是必须的" *women shi xuyao zhengfu zhichi de, zhe shi bixu de.*
3. *Fen* = cent.
4. On a scale going from 1 to 10, 1 meaning that people "completely disagree" and 10 meaning that people "completely agree", 73 percent of respondents answered 7 or above (World Value Survey, 2010–2014). As a comparison, only 65.9 percent American respondents answered 7 and above. A total of 23.6 percent Chinese respondents answered that they "completely agreed," compared to 12.6 percent American respondents.
5. Dragonhead enterprises are companies recognized by the Chinese government for their leading role in their industry sectors. The status grants them with certain tax exemptions and other financial support.
6. Expenditures allocated to Grain and edible oil reserves are not steadily increasing. Increase rates vary according to China's international supply strategy. We can indeed see, for instance, that the budget underwent a tremendous rise in 2009, just after the 2007–2008 international food price crisis (probably to replenish depleted stocks).

REFERENCES

Anderson, J. E. (2003). *Public policy-making: An introduction*. Boston: Houghton. Mifflin.

Blecher, M., & Shue, V. (1996). *Tethered deer: Government and economy in a Chinese county*. Stanford: Stanford University Press.

Boyd, R., & Ngo, T. W. (2005). *Asian states: Beyond the developmental perspective*. New York: Routledge.

Burns, J. P., & Zhou, Z. (2010). Performance management in the government of the People's Republic of China: Accountability and control in the implementation of public policy. *OECD Journal on Budgeting, 2*, 1–28.

Byrd, W. A., & Gelb, A. (1990). Why industrialize? The incentives for rural community government. In W. A. Byrd & Q. Lin (Eds.), *China's rural industry: Structure, development, and reform*. Oxford: Oxford University Press.

Chen, W. (1998). The political economy of rural industrialization in China: Village conglomerates in Shandong Province. *Modern China, 24*(1), 73–96.

Chien, S. S. (2010). Prefectures and prefecture-level cities: The political economy of administrative restructuring. In J. H. Chung & T. C. Lam (Eds.), *China's local administration: Traditions and changes in the sub-national hierarchy*. London/New York: Routledge.

Chung, J. H. (2010). Prefectures and prefecture-level cities: Political economy of administrative restructuring. In J. H. Chung & T. C. Lam (Eds.), *China's local administration: Traditions and changes in the sub-national hierarchy*. London/New York: Routledge.

Duckett, J. (1996). *The entrepreneurial state in China: Real estate and commerce departments in reform era Tianjin*. London: Routledge.

Edin, M. (2003). State capacity and local agent control in China: CCP cadre management from a township perspective. *The China Quarterly, 173*, 35–52.

Fan, H., Ye, Z., Zhao, W., Tian, H., Qi, Y., & Busch, L. (2009). Agriculture and food quality and safety certification agencies in four Chinese cities. *Food Control, 20*(7), 627–630. https://doi.org/10.1016/j.foodcont.2008.09.013.

Foster, K. W. (2002). Embedded within state agencies: Business associations in Yantai. *The China Journal, 47*, 41–65.

Gao, J. (2009). Governing by goals and numbers: A case study in the use of performance measurement to build state capacity in China. *Public Administration and Development, 29*, 21–31.

Göbel, C. (2011). Uneven policy implementation in rural China. *The China Journal, 65*, 53–76. https://doi.org/10.2307/25790557.

Guthrie, D. (1998). The declining significance of guanxi in China's economic transition. *The China Quarterly, 154*, 254–282.

Heberer, T., & Trappel, R. (2013). Evaluation processes, local cadres' behaviour and local development processes. *Journal of Contemporary China, 22*(84), 1048–1066. https://doi.org/10.1080/10670564.2013.795315.

Lampton, D. M. (1992). A plum for a peach: Bargaining, interest, and bureaucratic politics in China. In K. Lieberthal & L. GDM (Eds.), *Bureaucracy, politics and decision-making in post-Mao China*. Berkeley: University of California Press.

Landry, P. F. (2008). *Decentralized authoritarianism in China: The Communist Party's control of local elites in the post-Mao era*. Cambridge/New York: Cambridge University press.

Li, H., & Zhou, L. (2005). Political turnover and economic performance: The incentive role of personnel control in China. *Journal of Public Economics, 89,* 1743–1762.

Lieberthal, K. G. (1992). Introduction: The "fragmented authoritarianism" model and its limitations. In K. G. Lieberthal & L. GDM (Eds.), *Bureaucracy, politics and decision-making in post-Mao China.* Berkeley: University of California Press.

Lieberthal, K. G. (1997). China's governing system and its impact on environmental policy implementation. In F. Aaron (Ed.), *China environment series* (1st ed.). Washington, DC: The Woodrow Wilson Center.

National Bureau of Statistics Database. http://data.stats.gov.cn/workspace/index?m=hgnd

Nee, V., & Opper, S. (2012). *Capitalism from below: Markets and institutional change in China.* Cambridge, MA/London: Harvard University Press.

Oi, J. C. (1992). Fiscal reform and the economic foundations of local state corporatism in China. *World Politics, 45*(1), 99–126. https://doi.org/10.2307/2010520.

Oi, J. C., Singer Barbiaz, K., Zhang, L., Luo, R., & Rozelle, S. (2012). Shifting fiscal control to limit cadre power in China's townships and villages. *The China Quarterly, 211,* 649–675.

People's Journal. (2013, October 10). Notice about the improvement of the evaluation of leading cadres and leadership ranks of local party and government administration. *People's Journal* [关于改进地方党政领导班子和领导干部政绩考核工作的通知, 人民日报, 2013年12月10日 guanyu gaijin defang dangzheng lingdao banzi he lingdao ganbu zhengji kaohe gongzuo de tongzhi. *renmin   ribao*]   http://renshi.people.com.cn/n/2013/1210/c139617-23793409.html

Shen, C., Jin, J., & Zou, H. (2012). Fiscal decentralization in China: History, impact, challenges and next steps. *Annals of Economics and Finance, 13*(1), 1–51.

Smith, G. (2009). Political machinations in a rural county. *The China Journal, 62,* 29–59.

Walder, A. G. (1995). Local governments as industrial firms: An organizational analysis of China's transitional economy. *American Journal of Sociology, 101*(2), 263–301.

Wong, C. P. W. (2016). Interpreting rural industrial growth in the post-mao period. *Modern China, 14*(1), 3–30.

World Bank database. http://data.worldbank.org/

Zha, D., & Zhang, H. (2013). Food in China's international relations. *The Pacific Review, 26*(5), 455–479.

CHAPTER 5

Small Farmers "Endure or Escape"

Focusing on the forms of agrarian industrial entrepreneurship as was done in the previous chapters might lead the reader to reach the conclusion that the development of entrepreneurship, in the agricultural sector, is essentially taken care of by entrepreneurs not belonging to the social layer of farmers. On the opposite, I would like to underline that forms of agrarian capitalism have long existed among *nongmin* as well. The abolition of People's Communes and the implementation of the Household Responsibility System indeed enabled farmers to become independent in the decision-making linked to agricultural production at the beginning of the 1980s, pushing them to make farming choices according to market signals and to look for better profits. As such, small farmers can be considered as the first agricultural entrepreneurs. In addition, farming, in the end, is still mostly taken care of by *nongmin*, even though agri-food entrepreneurs, encouraged by local officials, took the leadership in agricultural modernization over the past few years.

What have *nongmin* entrepreneurs become? What is the place of the private entrepreneurship of small farmers in the contemporary process of agricultural modernization? Will the 300 million farmers be called upon to play a role such as happened in other countries through "entrepreneurs-paysans" (Muller 1984)? How do they react to current strategies implemented by local political and economic stakeholders? These are some questions this chapter would like to address.

© The Author(s) 2018
M.-H. Schwoob, *Food Security and the Modernisation Pathway in China*, Critical Studies of the Asia-Pacific,
https://doi.org/10.1007/978-3-319-65702-8_5

5.1   Institutional and Cultural Boundaries of the *Nongmin* Status

Reference to *suzhi* justifies social and political hierarchies of all sorts, with those of "high" quality gaining more income, power and status than the "low." In rural contexts, cadres justify their right to rule in terms of having a higher quality than the "peasants" around them. (Andrew Kipnis, *Suzhi: A Keyword Approach*)

5.1.1   Hukou and Land Tenure: Two Institutions Limiting Small Farmers' Ability to Become Farmers-Entrepreneurs

In the three decades following the abolition of collectivization, China underwent rapid urbanization. However, data show that in spite of the migration of a considerable population of rural dwellers to cities, the size of arable land per farmer remained small. The explanation of this situation can be found in the constraints that prevent migrants to transfer their land to farmers staying in the countryside. These constraints are rooted in two major institutional systems governing rural areas: the land tenure system and the *hukou* system. The property of rural land is in the hands of the Chinese state. It does not belong to farmers, who rent it to village committees. Since 2008, the Law on Land Contracts in Rural China grants farmers with rights over their land as if they owned it: they can sell, exchange, and inherit leases. However, in spite of this reform, permanent transfers of arable land are far from being common in rural areas.

The fact that land still belongs to local governments grants these latest with significant power over land transfer. Local officials have long preferred to favor entrepreneurs or real estate developers, as providing land to such players is likely to generate economic growth and to increase fiscal revenue. However, land requisition turned into a major source of conflict in rural areas (Yep 2013; Takeuchi 2013), pushing the central government to promulgate regulations to hinder arable land conversions. In 2008, the Ministry of Land and Resources set a red line of 1.8 billion mu, under which the total amount of arable land should not fall. Punishments of local cadres taking advantage of their rights over land at the expense of social stability became increasingly severe in the past few years, with the Ministry of Land and Resources recently warning local governments about the severity of the law regarding land use violations (Zhang 2014) and putting affairs on the public place and arable land conversion to nonagricultural purposes slowed down (Lin and Ho 2005).

The transfer of arable land, provided that it does not lead to the conversion of land to non-agricultural purposes, is strongly encouraged by the government as a way to increase the size of farms. However, the "farmland market" is far from efficient, as the current land leasing system and the *hukou* system create strong institutional obstacles hindering land consolidation. Since the beginning of the reform era, the *hukou* system underwent important changes. Restrictions on internal migrations disappeared, giving birth to a wide population of "migrant peasants-workers" (农民工 *nongmingong*), or former or temporary farmers working part-time or permanently in other sectors. Whereas the *hukou* system does not prevent rural–urban migrations anymore, it still keeps on separating the population into two categories: rural and urban dwellers. On *hukou* documents, two pieces of information (agricultural/nonagricultural work and place of residence) contribute to prevent rural migrants who live in urban areas to buy home, to have access to social security and retirement pension and to register their children in the public school system. In such a scheme, arable land replaces social security and retirement pension for migrant farmers who cannot have access to such services in urban areas. In order to be able to go back to farming in case of sickness, work injury, dismissal, or retirement, migrant workers usually leave their land to family members (e.g., parents) for free[1] or informally rent it to members of the extended family or to neighbors, sometimes for free (as can happen for low quality land), sometimes in exchange of a percentage of the harvest or in exchange of money. Informal land transfers are very common in rural areas. In the places where fieldwork was conducted, most of the land available for farming was cultivated, even though more than a third of villagers were working in the industrial sector, far away from the countryside. However, the number of permanent and official land transfers was limited in these areas. Land transfers did not appear on any official document and migrants could come back to farming whenever they wished or needed to.

In spite of repeated attempts to encourage cities to relax their *hukou* scheme, rigidities are still strong. The wish of the central government to reform the system bumps against the reluctance of provincial and municipal governments—especially in overpopulated cities of Eastern China—which claim that integrating migrant workers in urban social security, health, and education systems would have costs they would not be able to bear (Schwoob 2013a). As a consequence, countless small plots of land are still informally rented by a large population of former farmers not living in villages anymore, whether on a temporary or a permanent basis. This both distorts the picture given by national statistics—where the figures of informally rented farmland, sometimes on a long term basis, do not appear— and impedes the expected birth of a new category of "professional" farmers, cultivating secured pieces of land as their main business activity.

The land tenure system is currently undergoing major reforms as well. At the third plenum of the 18th Congress in November 2013, the land reform was a much-debated topic. According to the communiqué that was released after the plenum, the government wishes to "endow farmers with more property rights" (赋予农民更多财产权利, *fuyu nongmin gengduo caichan quanli*). Among other things, farmers, in a number of areas, are now able to transform their land into wealth in currency or other capital forms, such as loan collaterals (the absence of collaterals in rural areas being one of the main causes rooting farmers' difficulties to access credit (Schwoob 2013b)). However, guaranteeing farmers' rights in land transfers cannot be achieved without making their rights over land clearer. A tremendous amount of work is necessary to establish clear land rights, as in many areas, farmers still do not possess any certificate for their right to use land. Establishing a cadaster in rural areas requires collecting data on land use rights for dozens of millions of hectares of farmland, a task that promises to be arduous. In addition, establishing an official cadaster is likely to give rise to disagreements and conflicts, as local people will have to agree on land use rights on a permanent basis.

A number of local officials also expressed concerns about the land reform. According to them, giving land titles to farmers is likely to encourage them to take loans. Unable to reimburse loans, farmers would then lose their land and join the ranks of landless peasants, rooting more social uprise risks in rural areas, as was expressed by some of the interviewees. Tenuous progress has been made to reform the *hukou* and land tenure systems, but these reforms still face a strong reluctance of local governments to give up on economic and political power sources.

Local officials are not the sole opponents to reforms. Attempts to establish cadasters provoke vivid debates among farmers as well. In some villages I visited, I was told farmers were not satisfied with the current land allocation. According to an agent of the International Food Policy

Research Institute (IFPRI) coming back from fieldwork in Guangdong, farmers were afraid of a clarification of land use rights because of the imperfections of the current allocation system:

> Now the land reform is everywhere in China. But it is very complicated, because the farmers don't want to write down their plot. Actually, what happened is that in the 1980s, they were given 1 mu per person, but maybe this farmer got a less productive land, so on the paper, it is written that he only got 0.6 mus, and so it is unfair. (Interview, Beijing, December 2014)

In a village where I spent time in Anhui, farmers argued that households had evolved since the beginning of the 1980s, putting back into question the fairness of land distribution, even though reallocations were common when birth or death occurred within families. Conflicts would be likely to arise if official land titles would be set in stone.

In some places, farmers cultivate wider farms, thanks to informal land rental systems, which rapidly developed. In other places, arable land is subrented by farmers to entrepreneurs, who manage to gather large pieces of land and to develop "modern" farming on their own plots. Finally, in other regions, wide areas of land are left unfarmed. Because of the variety of situations and the informality of subrenting markets, the development of farming structures is difficult to follow, and it is almost impossible to assess the actual farm size with accuracy. However, drawing on fieldwork, the conclusion can be reached that accessing permanent and secure rights over a wide area of arable land is a challenge that is difficult to overcome, especially for small farmers.

5.1.2   The Importance of Cultural Schemes: The Rising Paradigm of Suzhi

In addition to the earlier mentioned institutional obstacles preventing farmers from escaping their social and economic condition—for instance, by acquiring more land and become farmers-entrepreneurs— one of the most striking things this research revealed is that the status of *nongmin* was associated with a strong negative connotation, deeply engraved in cultural schemes of *both* nonfarmers and farmers. *Nongmin* occupy the bottom rung of the socio-economic ladder, which partly explains why entrepreneurs hire farmers but never join their ranks—in the sense that they do not dare to grow products themselves. As was saying a manager in charge of DP projects in Jiangxi, who graduated from a CAU:

> Even if I had the opportunity to work in a farm and to live in the countryside, even if this is good for me, my parents will never accept that – and my grand-parents will even less accept it. How to say… They think that people don't respect people working in the countryside. It's not the same as in France, where people think that they can live a better life in the countryside some-times. Here, you live better lives in cities. (Interview, Jiangxi, October 2012)

Although the interviewee was then talking about rural dwellers in general, her statement is even truer for farmers. The term *nongmin*, in any case, usually encompasses both rural dwellers and farmers.

The low social status of *nongmin* is deeply engraved in cultural schemes. A wide corpus of literature developed on this topic and evidences that the status of Chinese farmers and rural dwellers is significantly lower than the one of the rest of the population. The research dealing with the notion of *suzhi*, or "population quality", is particularly enlightening on this topic (Anagnost 2004; Thogersen 2003; Murphy 2004; Kipnis 2006). The term *suzhi* started being used again in the 1980s, when the government started building discourses on development at the beginning of the reform and opening-up era. Particularly instructive is this quote from Murphy (2004: 177): "Suzhi derives part of its ideological potency through its reinforcement of related systems of valuation already embedded within Chinese development discourse, such as town versus country, developed versus backward, prosperous versus poor, civilized versus barbarian, and to have culture (you wenhua) versus to be without culture (mei wenhua)." According to the author, the categorization of groups within the population is in fact part of a political modernization program. As she says, "[…] although concerns about suzhi pertain to the entire population, groups in lower valued situations are seen to need special remedial attention. […] in a variety of social and historical contexts, nation-states perceive a problem in the 'backwardness' of certain groups, in this case rural people, and designate a pivotal role for schools in 'civilizing' them". In fact, the concept of *suzhi* has only been widening the divide between rural and urban dwellers, in the sense that rural dwellers and *nongmin* are seen as "low *suzhi*" or "low-quality" population. In rural areas, the program aiming at "Building a New Socialist Countryside"—which has been promoted by central officials and implemented by local officials since 2004—emphasizes the need to promote "urban and rural integration" (城乡一体化 *cheng-xiang yitihua*) and to "transform farmers into urbanites" (农民市民化 *nongmin shiminhua*) (Bray 2013), emphasizing again the superiority of the social status of urban dwellers compared to people from the countryside. According to Kipnis (2006), the word *suzhi* has now become central to the contemporary governance and society in China, in the way that reference to suzhi "justifies social and political hierarchies of all sorts, with those of 'high' suzhi being seen as deserving more income, power and status than those of 'low' suzhi". Suzhi has turned into a real paradigm almost legitimizing the low social status of certain groups within the Chinese society.

5.1.3   The Debate on Land Ownership

An increasing number of scholars have been advocating for land privatization, as a solution to social issues in rural areas (Prosterman et al. 1990; Sargeson 2012) and as a way to secure land rights and attract more people in the farming business. Zhang and Donaldson (2013), on the opposite, argue that "the participation of agribusiness in China's agriculture has helped to realize the central government's goal in reforming the agricultural sector", while the current system of collective land ownership would have provided farmers "with a tool to resist pressure from the companies", which would have had the result that "agricultural modernization in rural China has progressed in the more equitable ways described in these pages". Insights from fieldwork are far from corroborating the claim of Zhang and Donaldson. Class inequalities are still important between farmers (or farmers-workers) and entrepreneurs. The capacity of farmers to negotiate with powerful investors that is described by Zhang and Donaldson was nonexistent in the areas where I conducted fieldwork. The absence of private land ownership rights enables county and township governments to grab land in order to favor entrepreneurs. Farmers, deprived from a resource they could use to overcome the barriers of their social, cultural, and economic marginalization, face tremendous difficulties to access a stable status of "farmer-owner-entrepreneur" that could be attractive for future generations of farming labor force, and remain stuck at the social level of farmer or "semiproletarian" farmer-worker. Land privatization alone is unlikely to solve the issue, as this latest is also rooted in the institutional system—such as the *hukou* system—in cultural schemes and in established patterns of power in rural areas.

In addition, land ownership, in a number of developing countries, has proven to be detrimental to smallholders. As stated by Binswanger et al. (1995):

> Land rights and ownership tend to grow out of power relationships. Landowning groups have used coercion and distortions in land, labor, credit, and commodity markets to extract economic rents from the land, from peasants and workers, and most recently from urban consumer groups or taxpayers. Such rent-seeking activities reduce the efficiency of resource use, retard growth, and increase the poverty of the rural population.

As a consequence, the establishment of land ownership is generally considered as risky, given the current context of the Chinese countryside, not only by Chinese officials but also by a number of experts (Unger 2014).

5.2   RURAL MIGRATION: ESCAPING THE STATUS OF FARMER

5.2.1   *Leaving the Farming Sector*

Fieldwork demonstrated that farmers were not trying to put back into question their social status, deeply engraved in the cultural scheme of the nonrural society and also engraved in their own cultural scheme. *Nongmin* are aware of their low social condition, of their "low suzhi", which they do not put back into question but which they rather try to escape from. Young rural dwellers, in particular, wish to migrate to cities and/or to work in sectors other than farming. In most of the rural areas I went to, people between the age of 20 and 40 were missing. As was noting a farm manager in Jiangxi:

> There aren't young people anymore here. They all left to look for jobs (出去打工了 *chuqu dagong le*) in Guangzhou, Meizhou, everywhere [even if] conditions are very bad over there. (Interview, Jiangxi, October 2013)

Parents, on their side, also encourage their children to "look for better lives" in cities. Andrew Kipnis and a number of researchers reach similar conclusions. As Kipnis (2001) puts it: "The most obvious cause for rural educational discipline is a desire for social mobility. Throughout the reform era, Zoupingers have expressed this desire with the adage 'hoping one's child becomes a dragon' (wang zi cheng long)."

Farmers going out to look for jobs in cities or in the industrial sector are also attracted by the higher income that such lives promise them. Since the middle of the 1980s, China's economic growth mostly benefited urban households, who saw their revenue grow much more rapidly than rural households. Pushed away of the countryside by the difficulties they encounter as small farmers (both to increase the size of their land and to have access to credit), by the low status that they feel defines them, and by the development gaps between rural and urban areas, farmers usually adopt a going-out strategy. They are encouraged to do so by local governments and local urbanization targets and by enterprises of secondary and tertiary sectors, which see them as a convenient source of cheap labor.

Going-out is not just a way to access better economic conditions. It is also a way to escape one's social condition of *nongmin*, one's "low *suzhi*"— even if sometimes, even when they migrate to cities, farmers still consider themselves as "temporary and undesirable guests" (Froissart 2007: 217). Particularly enlightening is this quote from Kipnis (2001: 16–17): "The commitment to leave the countryside reflects not only the hope of relatively lucrative urban occupations but also, for many students, a desire to shed the stigma of the 'peasant' label." In fact, migrant farmers do not necessarily have better living conditions in cities, compared to farmers who stay in rural areas. Difficulties experienced by migrants living in urban areas with rural *hukous* are tremendous. Even if things are currently evolving (mainly in second and third tiers cities), many migrants still face restricted access to services such as health coverage or social security and have little hope of fulfilling locally set requirements to be granted urban *hukou*. Among the 700 million people living in urban areas, almost 230 million still hold a rural *hukou*. On the borderland of legality, this population is highly disadvantaged. Rural migrants generally do not enjoy a high level of education and are offered low wages and insecure and temporary jobs. In addition, they often work without employment contracts, which could be a first step toward pension rights, health insurance, and protection for workplace accidents, unemployment insurance, and family assistance. Only about half of the rural migrant population would have a fixed-term contract agreement, the rest being employed informally (Lan 2013). As Huang et al. (2012: 142) state it: "Those people generally take the heaviest and dirtiest jobs, are the most poorly paid, do not enjoy legal protections, and work without benefits or with reduced benefits."

However, rural dwellers, especially the young ones, are still willing to accept the tough conditions of nonfarming jobs to seize their chance to escape their *nongmin* status. Migration still remains their best option to get rid of the "peasant" stigma or the "low *suzhi*".

5.2.2   Going Out to Come Back as an Agricultural Entrepreneur

Going out can be a way for farmers to be freed from their social status and to come back as an entrepreneur. Whereas local farmers who had become entrepreneurs without ever leaving the countryside were rare and seen as people having achieved real miracles, former farmers having worked a certain amount of time in cities and having come back to launch businesses were more numerous. In Capitalism from Below, Nee and Opper (2012: 54) show that rural dwellers with modest origins, and especially farmers, significantly contributed to the rise of private entrepreneurship in China:

> Our Yangzi delta survey confirms that those who ventured into the private enterprise sector of the manufacturing economy came from modest to marginalized social backgrounds. The entrepreneurial movement was fueled neither by the technocratic elite of skilled engineers from state-owned companies nor by the country's political and administrative elite. [...] Although entrepreneurship is no longer exclusively a rural affair, rural founders are still prominent in the overall picture, with 53 percent of our respondents stemming from rural, and often farming, backgrounds.

Nee's and Opper's survey is mostly about private entrepreneurs having launched businesses in the manufacturing sector. A similar process of emerging capitalism started happening in the agricultural sector at the beginning of the 2000s, as a consequence of the new incentives given by local governments to entrepreneurs willing to engage in agriculture and food business. In the areas that I investigated for this research, there were two kinds of entrepreneurs investing in agriculture: local and nonlocal entrepreneurs. In "inland" areas such as Jiangxi, most of the agricultural entrepreneurs I met were local people. On the opposite, in Shandong, the origins of entrepreneurs were much more diverse, for several reasons, among which the local tradition of an export-oriented agriculture and a good business environment for both Chinese and foreign entrepreneurs.

The former occupational activity of entrepreneurs, on its side, varied a lot. Many "agricultural businessmen" I met used to work in completely other sectors and had no experience of farming whatsoever before they had engaged in agribusiness. As they had never been farmers, they did not have any land rights and needed to rent or to buy land use rights (使用 *shiyong*) from farmers, or to rent land directly from the government. In Jiangxi, many businessmen were renting land directly from the government, as in the beginning of the 2000s, local officials had decided to turn forests located on hilly areas into orchards suitable for citrus production. However, in reality, models of land usage are usually mixed. Businessmen who manage to rent a certain area of land directly from the government usually keep on relying on local farmers having their own land use rights, in order to raise production volumes. They rely on local farmers either through buying their products or subrenting their land.

Among the local entrepreneurs I met, a few had been farmers in the past. A number of entrepreneurs, in Lushan, for instance, were former farmer-migrants[2] who had spent a certain amount of time working in cities. However, a clear line was always drawn between "peasants" (农民 *nongmin*) and agricultural "businessmen" (生意人 *shangyiren* or sometimes 农场主 *nongchangzhu*). Figure 5.1 represents a drawing that was made by the manager of a small grocery store who was getting its products directly from farmers. The drawing clearly illustrates the difference between farmers (in red, owning use rights over small plots of land) and businessmen (in blue, getting their supplies both from their own plots and from the plots of small farmers).

Former farmers who managed to become businessmen had usually spent time working in cities or in sectors other than farming. The time spent in cities granted them with several kinds of capital: financial capital (as rural–urban gaps are still wide and, as a consequence, salaries are usually higher outside the farming sector) and also business-related knowledge as well as access to a network of contacts, likely to serve a variety of purposes (as noted by Ma and Cheng (2010: 906), who note the following functions of social networks in China: obtaining capital; securing information, raw materials, and technology; finding sales channels; and recruiting workers). When they come back to their hometown, former farmers benefit from this acquired capital, knowledge, and contacts as well as from their knowledge of the local area and of local networks, usually made of family ties. Being a "*bendiren*" (本地人 "native"), in some cases, facilitates the establishment of relationships with local officials and is useful to have access to land more easily.

Escaping rural life and "going to the city" is a way to be freed from the status of farmer and to be able to access another social status, which can

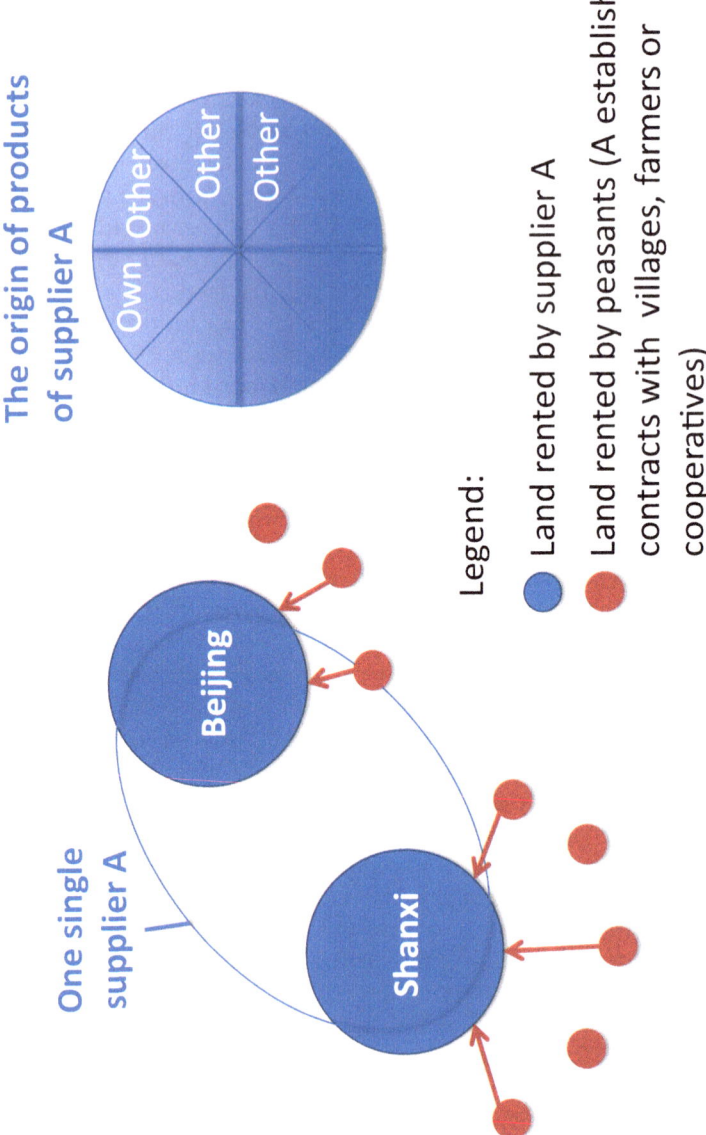

Fig. 5.1 "The origin of products of one supplier" (drawing made by the manager of a grocery store)

possibly lead to the status of "businessman"—even when these business-men actually engage in agricultural production. As was summing up the earlier mentioned manager of the grocery store:

> Some of the businessmen were farmers. This is one of the good things of the rapid development in China I think: people in their twenties can be farmers and then they go to live in the cities and they can become businessmen. (Interview, Beijing, November 2012)

Locked in an institutionally and culturally bounded social status, they can only escape by going out of the farming sector—even on a temporary basis—farmers seem to have been unable so far to take on a role in agricultural modernization. However, the recent push for the development of agricultural cooperatives once held out hope that the situation would evolve.

5.3   Agricultural Cooperatives Empowering Small Farmers?

> Step by step, the small and middle land ownership of the farmers, the basis of the whole political constitution, is succumbing to the competition of giant farms. […] This school of Socialism dissected with great acuteness the contradictions in the conditions of modern production. […] It proved, incontrovertibly, the disastrous effects of machinery and division of labor; the concentration of capital and land in a few hands; overproduction and crises; it pointed out the inevitable ruin of the petty bourgeois and peasant. (Karl Marx, Frederick Engels, *Manifesto of the Communist Party*)

5.3.1   *Farmers' Cooperatives Worldwide*

In a number of developed countries, farmers' cooperatives have been a useful way to mobilize farmers and to grant them with a role in agricultural modernization. The first agricultural cooperatives were founded in Europe at the end of the nineteenth century, as a response to low agricultural prices that severely impacted farmers in the 1880s and 1900s. The rationale of these new producers' groups was twofold. First, through joint purchasing, the members of cooperatives could have access to cheaper products (e.g., fertilizers and pesticides) and form a common pool of technological resources, recently made available by the Industrial Revolution.

At the same time, members of cooperatives, as a group, could improve their capacity to defend themselves against the abusive practices of input suppliers. However, the proportion of farmers belonging to agricultural cooperatives was remaining low at that time.

In the aftermath of the 1929 financial crisis, faced to the adverse consequences of an excessive laissez-faire capitalism, industrialized states started reasserting their role in the control of markets, and particularly in the control of agricultural markets. New support policies were created in order to supplement market mechanisms, which had proven insufficient to balance supply and demand. Agricultural cooperatives started being seen as efficient transmission belts for the new state-led agricultural development incentives, which included credit, insurance, and subsidies for basic agricultural inputs. For this reason, industrialized states started implementing legal and political environments suitable for the development of cooperatives as new corporatist groups. In the aftermath of the Second World War, their number rapidly increased. In France, cooperatives now represent 40 percent of the French agri-food sector, and three out of four farmers belong to at least one cooperative (Coop de France 2016).

Agricultural cooperatives are traditionally classified according to the three major functions they are meant to perform: supply, marketing, and services. Supply (or purchasing) cooperatives provide their members with affordable agricultural inputs, such as seeds, fertilizers, pesticides, fuel, and farm machinery. The basic principle of supply cooperatives is joint purchasing, which enables members to negotiate bulk prices. Marketing cooperatives support their members in selling products. The idea of marketing cooperatives is that farmers, as a group, have more bargaining power and can sell bigger volumes, which usually better meets the demand of modern retail. In addition, marketing cooperatives often raise the value-added products through vertical integration. Group investment makes it indeed possible for a given cooperative to purchase its own storage, processing and distribution infrastructures, and equipment—some cooperatives even have their own grocery stores—thus extending the control of farmers over markets. The last type of cooperatives provides its members with a wide variety of services, which would otherwise not be affordable to individual farmers. Such services may include information (e.g., trainings and consultancy), technical services (e.g., artificial insemination and herd management) or financial services (e.g., credit and insurance). Some cooperatives of services also provide access to electricity, communications, and even health care, schooling, and housing. In reality, most of agricultural cooperatives fulfill more than one of these three functions.

Usually, forms of cooperation in the agricultural sector are named "agricultural cooperatives" or "farmers' cooperatives" if they fulfill two criteria. First, cooperatives have to be member-owned enterprises, in the sense that each member is supposed to be an investor and to have stakes in the enterprise. The other criterion is that cooperatives have to be run on democratic principles, meaning that decisions regarding the strategy of the cooperative are taken through democratic vote or by representatives elected by its members through democratic vote (according to the international principles set by the International Co-operative Alliance, this also applies to nonagricultural cooperatives).

At the beginning of the twenty-first century, the movement of agricultural cooperatives has spread worldwide. However, their forms and the objectives they aim at fulfilling can differ widely according to countries. The objective of the following subsection will be to depict the peculiarities of the Chinese model of cooperatives.

5.3.2   *The Central Push for the Development of Chinese Cooperatives*

In China, agricultural cooperatives (合作社 *hezuoshe*) appeared for the first time in the 2005 Number One Document, in the subparagraph "Accelerate the building of circulation and examination infrastructures for agricultural products" of the paragraph "Strengthen the building of rural basic infrastructures and improve agricultural development environment", in the following sentence: "Seriously bring into play the action of supply and marketing cooperatives for (the improvement of) the circulation agricultural products, means of production, etc." In 2006, the law on cooperatives established a legal status for *nongmin hezuoshe* (农民合作社, farmers' cooperatives). According to the law, farmers' cooperatives shall be founded in rural areas by farmers. In addition, cooperatives are supposed to be created by member-farmers, who are supposed to be on an equal footing and to benefit from the services provided by the cooperative: "Farmers' cooperatives are established on the basis of rural households' contracted management, gathering service users and suppliers engaged in the production of a same kind of agricultural product, in a voluntarily contracted mutually beneficial economic association. The objective of farmers' cooperatives is to provide to its members: agricultural means of production, marketing, processing, storage and transport services, technology and information services, etc. Farmers' cooperatives should follow the below-listed principles:

- Member-farmers shall be the main body;
- Services provided to members shall serve the interest of all members;
- Members shall join the organization on a voluntary basis and be free to leave;
- Members shall be on an equal footing and democratic management shall be put into practice;
- Profits shall be redistributed to members according to their share in the cooperative."

From 2005 to 2013, central documents progressively granted an increasingly important role to farmers' cooperatives, calling local officials to support their development in order to speed up agricultural modernization. The number of occurrences of the word "cooperative" increased tremendously in Number One Documents from 2005 to 2013. While cooperatives were mentioned just once in 2005 and 2006 Number One Documents, there were no less than 28 references to them in the 2013 Number One Document (Fig. 5.2).

In addition, the role granted to cooperatives evolved and became much more diverse. In 2005 and 2006, only "supply and marketing cooperatives" (供销合作社 gongxiao hezuoshe) were mentioned in Number One Documents and their role appeared somehow limited to the improvement of the circulation of food products. In 2007, the role of supply and marketing cooperatives in the development of "modern rural circulation systems" (流通体系 liutong tixi) was mentioned again. In addition, "farmers' professional cooperatives" (农民专合作社 nongmin zhuan hezuoshe) emerged as "innovative" ways to promote systems and mechanisms that could enable the development of modern agriculture. The 2007 Number One Document encourages local governments to "do everything in their power" to promote the development of farmers' cooperatives and to support their efforts in the purchasing of means of production, in marketing, in information services, in technological training, in storage, and in the processing of agricultural products. In the 2008 Number One Document, alongside with supply and marketing cooperatives and farmers' specialized cooperatives, agricultural machinery cooperatives (农机合作社 nongji hezuoshe)

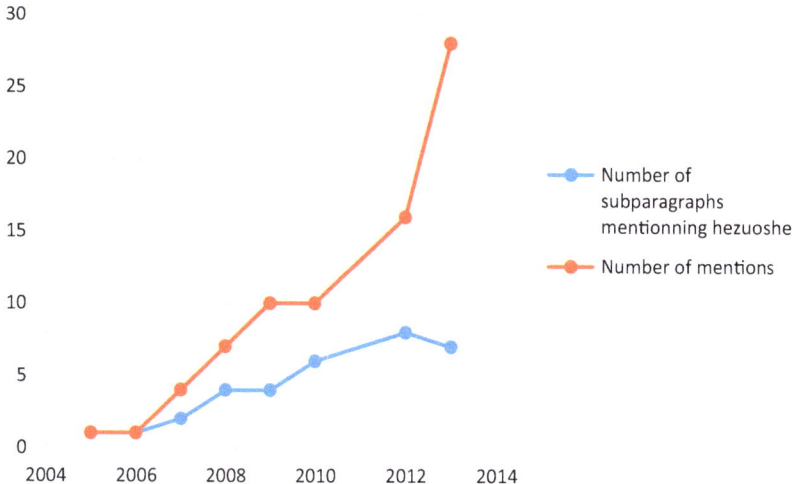

Fig. 5.2   An increasing focus on farmers' cooperatives in number one documents

are also mentioned. Farmers' specialized cooperatives, on their side, were from then on considered as important players in the development of agroindustrial capacities in rural areas (alongside with dragonhead enterprises) as well as new service providers to farmers (alongside with "rural service organizations" 农村服务组织 *nongcun fuwu zuzhi*). In 2009, 2010, 2012, and 2013, the role of farmers' cooperatives was further refined and extended to other services (e.g., financial services and purchase of improved seeds).

What is interesting to note is that 2010 onward, a subparagraph was added to Number One Documents, which underlines the necessity to improve Party building inside farmers' cooperatives. Could this be a sign of the success of agricultural cooperatives development policies? Data accurately quantifying the development of Chinese farmers' cooperatives are difficult to find. According to a survey made by Deng et al. (2010: 496), the effect of central policies promoting cooperatives was tremendous (Fig. 5.3). According to official statistics as well, the development of farmers' cooperatives was remarkable and the number of farmers' cooperatives would have reached 600,000 in 2012, gathering approximately 46 million farmers. Beyond these figures, what were the practical modalities of this development?

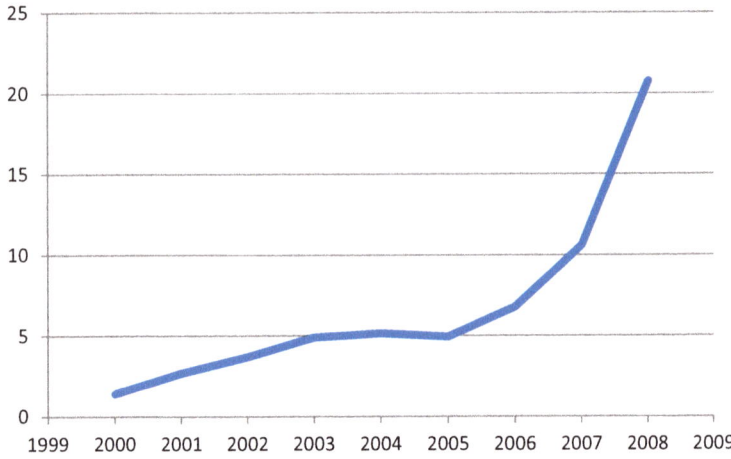

Fig. 5.3 Percentage of Chinese villages with professional farmers' cooperative (Source: Deng et al. 2010)

5.3.3 Modalities of Development: Insights from Fieldwork

When I conducted fieldwork in rural areas, I discovered that a number of farmers were already providing services which cooperatives were expected to provide, such as marketing, bulk purchasing of agricultural inputs, and lending of agricultural machinery. Farmers providing others with marketing services usually cultivate relatively wide areas of a designated product and own small trucks to reach other villages and township markets. They usually buy the yield of farmers living in their village and in surrounding areas (Fig. 5.4) and sell it to local wholesalers—either public or private (Fig. 5.5). It also happens that farmers buy agricultural equipment and rent it to other farmers. Usually, the formers are usually the ones who cultivate the biggest areas and have the financial resources to purchase equipments.

Informal forms of services such as marketing of agricultural products and renting of farm equipment have been common between farmers for a long time, outside the legal framework of farmers' cooperatives that was set by the central government in 2007. However, "farmers-merchants" were not always present in the areas that I explored, neither were they sufficient to cover the needs in terms of marketing and agricultural equipment. The existence of these proto-forms of exchange of services between

Fig. 5.4 Weighting the yield of small farmers in surrounding areas in Anhui (Photography by the author, Nov. 2014)

farmers thus does not put back into question the rationale of the recent governmental push for the development of farmers' cooperatives.

Farmers' cooperatives I investigated in Jiangxi usually gathered a variety of stakeholders much wider than the earlier discussed forms of agricultural cooperation—in which only small and middle-farmers take part. In fact, I could acknowledge the existence of two types of farmers' cooperatives. The first brings together only farmers—for instance, farmers who took the decision to market their products together. The second model involves the participation of industrial stakeholders based in rural areas as well.

During my fieldwork, I was often directed toward "the most modern" local agricultural structures—probably because I was explaining that I was conducting research on agricultural modernization and because local

Fig. 5.5 Selling bulk product in the township's purchasing bureau in Anhui (Photography by the author, Nov. 2014)

players were also probably eager to demonstrate their good achievements. In Jiangxi, the "most modern" farmers' cooperatives I investigated systematically included food transformation industries among their members. According to the Chinese law, industrial players can be shareholders in farmers' cooperatives. In Lushan, I discovered that local governments, who lack confidence in farmers' knowledge for the establishment of professional cooperatives, preferred to promote the "industry plus farmers" model of cooperatives. Concretely, most of the industrial players who were shareholders in farmers' cooperatives at the time I conducted fieldwork— usually food-processing enterprises based in rural areas—in fact already existed before the official status of agricultural cooperatives was enacted by the law in 2006. Created at the beginning of the 2000s, they existed even before the central government started promoting the development of cooperatives. In the aftermath of the promulgation of farmers' cooperatives by central documents, food-processing enterprises that already existed in rural areas were in fact encouraged to set up farmers' cooperatives. In Lushan, I was told that enterprises could easily take the identification documents of farmers working for them and register cooperatives in their name, as only farmers could register cooperatives—the only criterion being that at least five farmers shall take part in the project.

Insights from fieldwork in Jiangxi showed that rural-based food-processing enterprises registered farmers' cooperatives for several reasons. The first reason is linked to fiscal and financial support. Farmers' cooperatives can indeed benefit from targeted subsidies (e.g., for the purchase of agricultural equipment) as well as from tax abatements. If the terms and conditions of subsidies vary from one area to another, subsidies always exist for cooperatives and constitute an important incentive for enterprises to set such structures. Another reason mentioned by a number of interviewees was the maintenance of good relationships with local governments, who were particularly eager to promote the "mixed" model of farmers' cooperatives, as the following quote illustrates:

> The government thinks that maybe farmers are not well enough educated (they don't know how to use a computer, they cannot make invoice), and so they encourage local enterprises to set up cooperatives. (Interview with a manager of a DP project, Jiangxi, November 2012)

Local governments did not force entrepreneurs to create cooperatives, but rather opened new areas of opportunities (subsidies, easier procedures for license, etc.), which entrepreneurs were eager to grasp. Finally, I met factory managers who told me they had set up farmers' cooperatives to please their clients. Retailers can indeed be pushed by the government or by their own ideals to look preferentially for farmers' cooperatives to implement DP projects.

The legal status of farmers' cooperatives, which enables the formation of mixed models of shareholding, as well as the preference of local governments for the involvement of industrial players in the process, created unequal conditions for the development of the two types of farmers' cooperatives—the ones gathering only farmers and the ones gathering farmers as well as industrial players. However, this does neither mean that cooperatives gathering only farmers do not exist, nor that their emergence is not encouraged by local officials. However, when they reach a certain size, farmers' cooperatives usually evolve toward the model of "industry plus farmers" cooperatives. The manager of a cooperative I met in Jiangxi used to be a farmer and was encouraged to create a cooperative by the local government in 2009. The cooperative is now associated with a factory processing oranges that was established in 2006. Despite the fact that the cooperative was not created by the factory but by a former farmer and although elections of the manager of the cooperative and of the staff of the factory are held every three years, the division of responsibilities within the factory cooperative does not differ much from the one within other industrial–farmers cooperatives that I investigated. In these latest as well as in the cooperative created by a farmer, there is indeed a strong divide between the managers and the farmers-workers. The following quote—in which a manager of the factory explains the reasons why the former farmer does not farm anymore—illustrates how deep the divide is between the managers and the workers who actually grow products:

> The manager used to be a farmer, but he does not have time anymore to cultivate his orchard because of administrative work. So he lends his land to other farmers of the cooperative. (Interview, Jiangxi, October 2013)

To sum up, the fact that numerous farmers' cooperatives were created by agri-food industrial players is a particularity of the Chinese model that does not follow the principle of the dual status of shareholders that is often promoted elsewhere,[3] according to which shareholders shall be, at the same time, providers and beneficiaries of the services offered by the cooperative and should be put on an equal footing with the other members. Even when the principles of democratic management promoted by Chinese central documents are effectively enforced, members of farmers' cooperatives do not appear to be on an equal footing, as managers and decision-makers usually put an end to their farming activities and leave their *nongmin* status behind. Researchers having investigated different areas in China have reached quite similar conclusions. For instance, Yan and Chen (2013: 969) state that "the contemporary support for the cooperative movement is confronted with the predominance of 'fake cooperatives', in which small producers barely participate". Further in the article, the authors quote a study conducted by Liu Laoshi, according to which "among the 272,000 cooperatives formally registered by 2010 in China, it is estimated by many observers that 80–95 percent of them are fake".

Other research states that cooperatives can be headed not only by enterprises but also by village committees or government departments as well (Zhang and Zhang 2007). As a consequence, in spite of the recent efforts of the government to promote agricultural cooperatives, the role of farmers remains small in agricultural modernization. Could the Community-Supported Agriculture (CSA) that is emerging in the suburban areas of big cities like Beijing constitute another alternative pathway to the widespread model of industrial players?

5.4 Community-Supported Agriculture as an Alternative Model?

Recent years have seen the emergence of a network of "green" producers launching CSA projects in rural areas surrounding cities such as Beijing, Shanghai, Chongqing, and Xiamen. This rapidly spreading movement finds its rationale in the growing concerns linked to food safety, fuelled by the scandals that are regularly brought to light by the media. The food challenges that Beijing experiences are numerous, both in terms of food security and food safety. Beijing's huge urban population—almost 20 million according to the last national official survey (2010)—needs tremendous amounts of food, which cannot be produced by the sole resources of the administrative area of the municipality. Land is scarce because of the continuous extension of urban areas, which drives prices up. In addition, water resources are low and much demanded to answer the needs of the urban population. According to the World Bank, water availability in the Hai River basin, which includes Beijing and Tianjin, was only of 300 cubic meters per person in 2007 (Xie et al. 2009: 1), way below the absolute water scarcity limit fixed by the FAO (500 cubic meters per year per capita). Agriculture, as it only contributes to about 1 percent of the GDP of the municipality, is usually not considered as a priority (interview with a director of the Beijing bureau of agricultural technology extension services, November 2013). As a consequence, only 30 percent of the food consumed by Beijing citizens comes from surrounding areas.

In addition to food security issues, challenges in terms of food safety are also important. As about 70 percent of the food comes from outside of the municipality through long food chains, a lot of citizens lack confidence in the origin of products. The concerns of Beijing's citizens in terms of food safety are particularly high. This awareness of food safety issues, added to the geographical proximity of the population to ruling authorities and to the wealth of citizens,[4] provides a suitable ground for the development of alternative farming models and smaller food chains, such as CSA.

The first forms of CSA were established in Japan. Within associations known as *teikei* (提携), Japanese consumers buy agricultural products directly from farmers through a system of annual subscription. *Teikei* were born in the 1960s and originate from the growing concerns of urban mothers worrying about the safety of food products (Amemiya 2011: 44). The movement really started spreading throughout the country after the creation of the Japan Organic Agriculture Association in 1971, which promoted organic agriculture as well as the principle of consumers–farmers relationships (Henderson and Van En 1999). *Teikei* were providing answers to the rise in consumers' food scares following the industrialization of farming and the worsening of environmental pollution,[5] which raised the demand for domestically produced organic food. In Western Europe, similar movements started emerging in parallel with the development of *teikei* in Japan in the 1970s and 1980s. In the middle of the 1980s, the concept spread from Switzerland and Germany to the United States with the two first CSA farms simultaneously created in 1986 in Massachusserts (the "Indian Line Farm") and New Hampshire (the "Temple-Wilton Community Farm") (Henderson and Van En 1999). Today, the concept has expanded worldwide and forms of CSA now exist in North and South America, Australia, Africa, Asia, and Western and Eastern Europe.

CSA farms, throughout the world, share common principles. First, within CSA farms, consumers purchase a membership and, in return receive seasonal agricultural products (usually a box of vegetables on a weekly basis). Such a system enables farmers to receive payment early in the season, which considerably improves their cash flow. Logistics is also eased, as the composition of boxes of vegetables is generally set by farmers, who do not have to cope with the variability in consumers' demand. Consumers, on their side, enjoy weekly supply of locally produced fresh organic vegetables at affordable prices with the insurance of knowing where, how, and by whom they were cultivated. Another important principle of the CSA model is risk sharing. In most CSA farms, members pay in advance and farmers do their best to provide them with an abundant box of products each week. In case of poor harvest due to unfavorable weather or pests, members are generally not supposed to be reimbursed. Consumers contracting membership in CSA farms thus are usually not only motivated by the quality of products but are also engaged in supporting local farming. As stated by Cooley and Lass (1998: 229): "Quality of produce was cited by 93 percent of the members surveyed as an important reason for joining a CSA. […] Support for local farming was also an important factor for 97 percent of the CSA members surveyed." Differences exist between CSA farms worldwide. They are linked to the way of sharing the farm budget (the duration of the engagement of consumers generally varies from one month to one year) and to the way of delivering products (e.g., French CSA farms—Associations pour le Maintien d'une Agriculture Paysanne [Association for the preservation of peasant agriculture] or AMAP—usually have a delivery points [farm, grocery store, school, enterprise, etc.] and set the day and hour for the picking up of vegetables by consumers).

The movement recently expanded to China. The first Chinese CSA farm was founded in 2008 by a Chinese scholar, following years of research on rural development in China as a graduate student and a PhD candidate at the Renmin University and a few months spent in a CSA farm in the United States. Supported by academic and political networks, the project rapidly proved successful. Land was easily acquired in the suburban rural areas of the municipality and regular conferences held on the topic helped to promote the model within the urban community of Beijing consumers. Four years after its creation, the "Little Donkey Farm" already enjoyed the support of more than 1000 regular clients. The success of the farm encouraged numerous investors to launch similar projects. There are today dozens of CSA-like farms around Beijing and their number has reached 200 throughout the whole country (Fig. 5.6).

In Beijing, the managers of CSA farms who I interviewed between mid-2012 and the end of 2013 frequently complained about the lack of interest of local officials in their projects and in agriculture in general. They claimed that they were experiencing financial difficulties and were not supported enough by the government—subsidies targeting essentially farmers. To a certain extent, CSA farms indeed suffer from a lack of interest of the government of Beijing, where agriculture accounts for a tiny share of the GDP and where officials are much more concerned about the challenges brought by urbanization in terms of pollution, transportation systems, electricity infrastructures, and migrations. However, I met a certain number of government officials interested in CSA projects and willing to promote them. In fact, the political power is fragmented among an important number of cadres, who can decide whether to support or not CSA

Fig. 5.6 Indoor farmers' market gathering CSA farms inside a mall in Beijing (Photography by the author, April 2013)

projects for a wide number of reasons that sometimes have nothing to do with the context but much rather deal with their personal interests and concerns, personality, and energy. As this quote from a manager of a CSA farm in Beijing illustrates it:

> At the beginning, we did not tell the township authorities about our project here, and we started conducting the project without noticing them about it. Of course, the village committee knew about it. But not the township government. We haven't told the government for several months. Eventually they found out about us. And the leader of agricultural development of the township government started to be very interested in our project, because he is young and he really wants to do something. At the beginning of each year we have to report to him about our activity. In fact, everything really depends on individuals. It depends on personality. (Interview, Beijing, April 2013)

In addition, it would be unfair to say that officials do not care about CSA projects, as debates do exist in academic circles (the profile of the scholar and first founder of CSA proves it) as well as inside local bureaus such as the agricultural technology extension center. On the picture of the front cover of the review that is published by Beijing's agricultural bureau and agricultural extension services, we can read: "社区支持农业(CSA):城市居民参与农业的新模式" (shequ zhichi nongye (CSA): chengshi jumin can yu nongye de xin moshi), or: "CSA: the participation of urban citizens and new agricultural models," proving that CSA are a matter of interest for Beijing's officials (see Fig. 5.7).

However, the recent mushrooming of CSA farms and the fact that people talk about CSA within governmental and academic circles do not necessarily mean that the model is scaling up throughout the whole country. The development of Chinese CSA is in fact limited by its characteristics, which are slightly different from the kind of CSA that spread in other countries. CSA farms I visited around Beijing usually operate according to the principle of membership for consumers. These latest pay in advance for boxes of fresh vegetables, to which are sometimes added fruits and eggs. In that, Beijing CSA farms do not differ much from their American, European, or Japanese counterparts. Chinese CSA farms usually deliver their products to farmers' markets (about five farmers' markets were operated in Beijing at the time fieldwork was conducted) or directly to the home of consumers.

The principle of risk sharing, however, was not enforced by the CSA farms that I investigated in the same way it is usually enforced elsewhere. Although consumers do pay in advance for boxes of vegetables, managers of CSA farms told me that many consumers often wished to change the composition of boxes of vegetables according to their weekly or daily needs. Calls, text messages, or e-mails—sometimes on the eve of the delivery—expressing requests for the composition of boxes (including demands for certain types and/or quantities of vegetables) were frequent. The price paid by consumers-members is in fact tremendous compared to the price of vegetables that can be purchased in supermarkets. On Fig. 5.8, we can see that the price for 12 to 32 boxes of vegetables (of 8 to 10 jin[6]), weekly delivered, goes from 1440 to 3584 RMB (approximately €170 and €430) (see also Fig. 5.9). Considering that the price of leafy vegetables and cucumbers of the species that were usually put in such boxes is around 2 to 3 RMB per jin on traditional markets, it means that the price of the boxes sold by CSA farms is five to seven times the one of usual vegetables.

In China, CSA farms have in fact built their success on the marketing of "safe" products, which became popular among middle-class urban consumers worrying about their health—even though most of the CSA farms I investigated were not selling food labeled as "organic", mostly because

Fig. 5.7   Front cover of the review 北京农业 *Beijing nongye* [Beijing's agriculture], August 2013

套餐种类	每周菜量	配送频率	配送次数	份额菜金	送菜方式	月均消费
两人份套餐	约8斤		32次	3584元		448元
三人份套餐	约10斤/次		32次	4160元	免费配送到家	520元
体验套餐	约8斤/次	每周1次	12次	1440元		480元
鸡蛋套餐	10枚/次		32次	800元		100元
柴鸡套餐	1只/月		8次	960元	需与蔬菜一起配送	120元

Fig. 5.8   Prices for 12–32 boxes of vegetables (8–10 jin), weekly delivered: 1440–3584 RMB (Photography of a flyer of a Beijing CSA farm)

2013年常季配送方案

家庭人口	每周菜量	配送频率	送菜次数	送菜总量	菜金	配送费	合计
二人家庭	8斤	一周一次	30次	240斤	1920元	450元	2370元
三人家庭	12斤	一周一次	30次	360斤	2880元	450元	3330元
四人家庭	16斤	一周两次	60次	480斤	3840元	900元	4740元
五人家庭	20斤	一周两次	60次	600斤	4800元	900元	5700元

2013年冬季配送方案

家庭人口	每周菜量	配送频率	送菜次数	送菜总量	菜金	配送费	合计
二人家庭	8斤	一周一次	22次	176斤	2112元	450元	2562元
三人家庭	12斤	一周一次	22次	264斤	3168元	450元	3618元
四人家庭	16斤	一周两次	44次	352斤	4224元	900元	5124元
五人家庭	20斤	一周两次	44次	440斤	5280元	900元	6180元

Fig. 5.9   Other tables of prices for 22–60 boxes of vegetables (8–20 jin), delivered on a weekly basis or twice a week: 2370–6180 RMB [between 280 and 740 euros] (Photography of a flyer of a CSA farm)

of the expensive price of labels and because of the lack of trust of consumers in organic labels. These consumers, even though offered with the possibility to visit CSA farms, usually stay remote from the realities of farming and expect a high level of service accordingly to the high amount of money they pay to get safe products delivered to their home. Receiving lower quantities of vegetables in case of bad weather conditions or pests is thus hardly acceptable for Chinese members of CSA farms. The products that are sold by CSA farms are "luxury" products. Chinese CSA is thus closer to a model of organic farms delivering safe products to a wealthy clientele contracting membership primarily for health purposes than to the model of CSA farms operating thanks to the involvement of consumers-members willing to share risks and to preserve local agriculture and its environment. The fact that CSA food is considered as a luxury good does not help the model to scale up, as it currently addresses a limited category of the population made of well-off residents of big cities worrying about their health. For most of Chinese citizens, who still dedicate one-third of their expenditure to food, buying vegetables more than five times more expensive than in traditional markets is simply not an option.

A final characteristic of the Chinese CSA farms that I investigated was the fragmentation of tasks. While only farmers (*nongmin*) were growing products, other tasks such as marketing, packing, and delivering were performed by other people—usually the founder of the farm along with a recruited team of people coming from the city. None of the Chinese CSA farms investigated had been created by former farmers. The sociological profile of founders of CSA farms in fact varied little. Founders had usually spent several years working outside of the farming sector, in various fields, from industry to energy, real estate, hotel business, and so forth. Having worked for several years in lucrative sectors, they had managed to save money and to establish the necessary contacts to launch a business. Some—in particular, the ones having worked in hotels in rural areas—had also established useful contacts in rural areas.

One of the main motivations mentioned by founders was the desire to bring solutions to their own concerns in terms of food safety. Most of them told me that they had started seriously worrying about their health or the health of their newly born children, and that this constituted the triggering factor for their decision to launch a business in green agriculture. In addition to the wish to bring solutions to an issue directly affecting them, founders of CSA farms were behaving as rational economic actors as well, seeing in the development of green food production bases an opportunity to create their own enterprise and to make profits.

Chinese CSA farms rely on farming labor to grow products, usually made of unskilled local farmers—in the sense that they did not have any vocational training in their lives, not that they did not know how to grow crops. Managers of CSA farms expressed difficulty in finding farmers to develop their activity or to replace labor on the eve of retirement. However, they were barely questioning the principle of task fragmentation, a principle they were justifying by the fact that "farmers are not good at writing or keeping data" (Interview, Beijing, April 2013). CSA farms thus do not completely eliminate intermediaries between producers (farmers) and consumers, and pose social problems similar to the model of investors—in particular, the question of the replacement of retired workforce—still not granting small farmers with the possibility to take part actively in agricultural modernization.

5.5   Conclusion

The dominant mode of action for the implementation of agricultural modernization—local governments relying on rural food processing enterprises to modernize farmers—is not completely preventing other models to develop. China remains a decentralized country, in which the state operates through a network of government officials, who are, above all, rational and individual actors. Because of the fragmentation of the state and because of the diversity of the Chinese territory, "pockets" of innovation exist, that dedicate more important efforts to the solving of social and environmental issues. However, these models (agricultural cooperatives or CSA farms) do not fundamentally put back into question the dominant frame of reference for agricultural modernization. While agricultural cooperatives usually do not change anything in local patterns of power between farmers and entrepreneurs, innovation born near Beijing is not fundamentally providing any alternative solution to the marginalization of *nongmin* that is acknowledged elsewhere in the country.

In addition, the spreading of alternative innovative models such as CSA farms is limited by a number of factors linked to the particularities of these models and of the cultural and social factors. Launching a business linked to organic food is indeed a behavior limited to the wealthiest and most environmentally conscious people in big cities and are probably not likely to spread to other layers of the Chinese society. For the rest of the Chinese consumers, strategies to curb food safety issues remain limited. Organic products are still unaffordable to the vast majority of consumers, considering the fact that the price of organic vegetables is more than five times the price of conventional products, and that Chinese households still spend more than the third of their expenditure in food. In addition, the booming organic food sector is not tightly enough controlled by the authorities and still lacks credibility (Li et al. 2011). At the exception of the wealthiest consumers and of a part of farmers who grow their own organic food, organic products are not an option yet to solve the issue of food safety—and, as a consequence, the issue of environmental degradation.

The marginalization of small farmers in the modernization process is detrimental to this category of the population considered to have a "low *suzhi*". It also has an adverse impact on the degree of environmental and social sustainability of the pathway China's agriculture is embarking on, as we will see in the last chapter.

NOTES

1. As a manager of a farm was telling me in Jiangxi: "[People who went to look out for jobs in cities] do not rent their land to other farmers, because their parents are here. In Jiangxi, people have children very early, many people get married around 18 years old. It means that their parents are about 40 years old, and still young" (Interview, Jiangxi, October 2013).
2. In Lanshui, the origins of entrepreneurs were much more diverse.
3. In France, the legal status of cooperatives is defined by the Code Rural, which states that members are "associés coopérateurs", meaning that they are, at the same time, users of the services provided by the cooperative and associates (investors) of the cooperative (Art. L521-3). As a consequence, only farmers can be "associés coopérateurs". Charters of cooperatives can stipulate that they may admit "associés non coopérateurs" (who can be non-farmers), but the status and advantages of "associés non coopérateurs" is strictly delineated by the law, for their share in the cooperative's capital (which cannot exceed 20 percent) (Article L522-3), their return in capital (Article L522-4) as well as their representativeness in the general assembly (they cannot hold more than one fifth of the votes) (Article L522–4).
4. Beijing is one of the wealthiest provinces in terms of net income per capita in urban areas, with 41,103.1 RMB in 2012 according to the National Bureau of Statistics of China (only Shanghai exceeds this figure, with 44,754.5 RMB).
5. Among which the Minamata disease: the discharge of methyl mercury by a chemical factory in Minamata bay between the 1930s and the 1960s caused the death of nearly 1000 people and affected several thousand people with strong neurological syndromes resulting from mercury poisoning following the eating of shellfish and fish.
6. 1 jin = half a kilogram.

REFERENCES

Amemiya, H. (2011). Genèse du Teikei: Organisations et groupes de jeunes mères citadines. In H. Amemiya (Ed.), *Du Teikei aux AMAP: Le renouveau de la vente directe de produits fermiers locaux*. Rennes: Presses universitaires de Rennes.

Anagnost, A. (2004). The corporeal politics of quality (suzhi). *Public Culture, 16*(2), 189–208. https://doi.org/10.1215/08992363-16-2-189.

Binswanger, H. P., Deininger, K., & Feder, G. (1995). Power, distortions, revolt and reform in agricultural land relations. In *Handbook of development economics*. Washington, DC: World Bank.

Bray, D. (2013). Urban planning goes rural. *China Perspectives, 3*, 53–62.

Cooley, J. P., & Lass, D. A. (1998). Consumer benefits from community supported agriculture membership. *Review of Agricultural Economics, 20*(1), 227–237. https://doi.org/10.2307/1349547.

Coop de France. (2016). http://www.coopdefrance.coop/fr/16/une-reussite-economique-et-sociale/

Deng, H., Huang, J., Xu, Z., & Rozelle, S. (2010). Policy support and emerging farmer professional cooperatives in rural China. *China Economic Review, 21*, 495–507. https://doi.org/10.1016/j.chieco.2010.04.009.

Froissart, C. (2007). Quelle citoyenneté pour les travailleurs migrants en République Populaire de Chine ?: l'expérience de Chengdu. Thèse : Sciences Politiques : Paris : Institut d'Etudes Politiques.

Henderson, M., & Van En, R. (1999). *Sharing the harvest: A guide to community supported agriculture*. White River Junction: Chelsea Green Publishing.

Huang, P. C., Yuan, Y., & Peng, Y. (2012). Capitalization without Proletarianization in China's agricultural development. *Modern China, 38*(2), 139–173. https://doi.org/10.1177/0097700411435620.

Kipnis, A. (2001). The disturbing educational discipline of "peasants". *The China Journal, 46*, 1–24. https://doi.org/10.2307/3182305.

Kipnis, A. (2006). Suzhi: A keyword approach. *The China Quarterly, 186*, 295–313.

Lan, F. (2013). The gap between urban and rural areas: the barriers to the dismantlement of transformation of rural migrants into urban residents, Caixin New Century 1, 7 January 2013 [蓝方, 城乡鸿沟:农民市民化中那些待拆的壁垒, 财新《新世纪》, 07/01/2013 *Lan Fang, chengxiang honggou: nongmin shimin hua zhong naxie dai chai de bilei, Cai xin "Xin shiji"*].

Li, J., Meng, X., & Liu, C. (2011). Saying "I love you" to organic products is not easy. Tianjin Network, 31 Oct 2011 [李家宇, 孟兴, 刘畅, "有机食品,想说爱你不容易", 天津网, 31/10/2011 *Li Jiayu, Meng Xing, Liu Chang, Youji liangshi xiang shuo ai ni bu rongyu, Tianjin Wang*] http://www.tianjinwe.com/tianjin/tjcj/201110/t20111031_4495764.html

Lin, G. C. S., & Ho, S. P. S. (2005). The state, land system, and land development processes in contemporary China. *Annals of the Association of American Geographers, 95*(2), 411–436.

Ma, W., & Cheng, J. Y. S. (2010). The evolution of entrepreneurs' social networks in China: Patterns and significance. *Journal of Contemporary China, 19*(67), 891–911. https://doi.org/10.1080/10670564.2010.508590.

Muller, P. (1984). *Le technocrate et le paysan: essai sur la politique française de modernisation de l'agriculture: de 1945 à nos jours.* Paris: Ed. Ouvrières.

Murphy, R. (2004). Turning peasants into modern Chinese citizens: 'population quality' discourse, demographic transition and primary education. *The China Quarterly, 177*, 1–20.

Nee, V., & Opper, S. (2012). *Capitalism from below: Markets and institutional change in China.* Cambridge, MA/London: Harvard University Press.

Prosterman, R. L., Temple, M. N., & Hanstad, T. (1990). China: A fieldwork-based appraisal of the household responsibility system. In R. L. Prosterman (Ed.), *Agrarian reform and grassroots development* (pp. 103–138). Boulder: Rienner.

Sargeson, S. S. (2012). Villains, victims and aspiring proprietors: Framing 'land-losing villagers' in China's strategies of accumulation. *Journal of Contemporary China, 21*(77), 757–777.

Schwoob, M. H. (2013a). L'intégration des immigrés de l'intérieur. *China Analysis, 42*, 8–12.

Schwoob, M. H. (2013b). La réforme de la finance rurale. *China Analysis, 46*, 38–42.

Takeuchi, H. (2013). Survival strategies of township governments in rural China: From predatory taxation to land trade. *Journal of Contemporary China, 22*(83), 755–772. https://doi.org/10.1080/10670564.2013.782125.

Thogersen, S. (2003). Parasites or civilizers: The legitimacy of the Chinese Communist Party in rural areas. *China: An International Journal, 1*(2), 220–223. https://doi.org/10.1353/chn.2005.0038.

Unger, J. (2014). The third plenum and rural property rights: Decisions in the right direction. In P. Harris (Ed.), *China at the crossroads: What the third plenum means for China, New Zealand and the world.* Victoria: Victoria University Press.

Xie, J., Liebenthal, A., Warford, J. J., Dixon, J. A., Wang, M., Gao, S., Wang, S., Jiang, Y., & Ma, Z. (2009). *Addressing China's water scarcity recommendations for selected water resource management issue.* Washington, DC: The International Bank for Reconstruction and Development/The World Bank.

Yan, H., & Chen, Y. (2013). Debating the rural cooperative movement in China, the past and the present. *Journal of Peasant Studies, 40*(6), 955–981. https://doi.org/10.1080/03066150.2013.866555.

Yep, R. (2013). Containing land grabs: A misguided response to rural conflicts over land. *Journal of Contemporary China, 22*(80), 273–291. https://doi.org/10.1080/10670564.2012.734082.

Zhang, D. (2014). Going a step further in impartial fair honest and efficient enforcement of the law. Published on the website of the Ministry of Land and Resources of China, 21 May 2014 [张德霖,进一步严格公正廉洁效能执法, 中国国土资源部网, 21/05/2014 *Zhang Delin, jinyibu yange gongzheng lianjie xiaoneng zhifa, zhongguo guotu ziyuan bao*] http://www.mlr.gov.cn/xwdt/jrxw/201405/t20140521_1317621.htm

Zhang, Q. F., & Donaldson, J. A. (2013). China's agrarian reform and the privatization of land: A contrarian view. *Journal of Contemporary China, 22*(80), 255–272.

Zhang, K., & Zhang, Q. (2007). Puzzles and thoughts about the growth of farmers' professional cooperatives. *Agricultural Economic Issues* (5):62–70 (*Nongmin zhuanye hezuoshe chengzhang de kunhuo he sikao, Nongye jingji wenti*).

Resulting Lock-Ins Impeding Transition Toward Environmental and Social Sustainability

6.1 A Rising Political Willingness to Answer Food Safety and Environmental Issues

Environmental protection and the safety of food products are closely linked. A great number of food safety issues are indeed caused by unsuitable farming practices. Among these practices, the spreading of pesticides and herbicides damages ecosystems and at the same time leads to residues on food products that are highly detrimental to human health. Food safety has been a recurrent theme of Number One Documents since 2004. It can be found in seven out of the ten documents as the subject of a whole subparagraph. Most of the policy guidelines linked to this topic promote better control and regulation of markets. During the years following the 2008 melamine milk scandal, which caused the sickness of hundreds of thousands of babies and killed several, policy guidelines became more precise and more pressing.

Environmental concerns, on their side, were already part of Number One Documents starting from the middle of the 2000s. Although policy guidelines were rather vague at the beginning, concepts were progressively refined along the years. In 2005, environmental concerns were linked to the necessity to protect water resources and to improve the resilience of agriculture to natural disasters. Then, new terms such as "circular agriculture", clean energy, and biomass gradually emerged (Table 6.1).

© The Author(s) 2018
M.-H. Schwoob, *Food Security and the Modernisation Pathway in China*, Critical Studies of the Asia-Pacific,
https://doi.org/10.1007/978-3-319-65702-8_6

Table 6.1 Emphasis put by 2004–2014 Number One Documents on land protection, ecology and food safety (occurrences in paragraph titles and subtitles)

	Soil quantity/quality	Promote ecology	Improve the quality and safety of food products
2004	*Not mentioned neither titles nor in subtitles*	*Not mentioned neither titles nor in subtitles*	2.a) "Raise quality and safety levels of food products"
2005	2. "Build a rigorous system protecting cultivated areas and improve the quality of cultivated areas"	3. "Reinforce irrigation and water conservancy and ecology and raise agriculture's capacity to cope with natural disasters"	*Not mentioned neither titles nor in subtitles*
2006	4.a) "Improve irrigation and water conservancy, farmland quality and ecology"	2.f) "Accelerate the development of circular agriculture"	*Not mentioned neither titles nor in subtitles*
2007	2.b) "Improve soil quality"	2.a) "Improve irrigation and water conservancy" 2.c) "Accelerate the development of clean energy in rural areas" 2.e) "Raise rural areas' sustainable development capacity"	5.b) "Improve market services and management of quality and safety of products capacity"
2008	3.d) "Strengthen the protection of soil and improve soil quality"	3.f) "Continuously promote ecology"	2.c) "Strengthen the standardization of food products and improve the quality and safety of products"
2009	4.c) "Strictly enforce cultivated land protection and land saving mechanisms"	3.e) "Promote ecology as a priority"	2.d) "Strictly regulate and control the quality and safety of food products"
2010	*Not mentioned neither titles nor in subtitles*	2.f) "Build a strong ecological security barrier"	*Not mentioned neither titles nor in subtitles*
2012	*Not mentioned neither titles nor in subtitles*	5.d) "Strengthen ecological construction"	6.c) "Improve agricultural products regulation and control"
2013	*Not mentioned neither titles nor in subtitles*	6.d) "Encourage the construction of rural ecological civilization"	1.d) "Enhance agricultural markets' regulation and control" 1.e) "Promote food safety"
2014	*Not mentioned neither in titles nor in subtitles*	3. "Establish sustainable agricultural development long-term mechanisms"	1.c) "Strengthen agricultural markets' control systems" 1.e) "Closely watch over the quality and safety of food products"

6.2 THE EFFECT OF THE PRODUCTIVIST FRAMEWORK ON BIODIVERSITY AND RESOURCES

At the local level, environmental protection and food safety guidelines, although stated in central documents early in the 2000s, were not of equal importance with the ones linked to food security or to rural living standards. In areas investigated in Jiangxi and Shandong, environmental issues were clearly second-rank objectives for local officials. In fact, most of the environment-friendly farming practices were put aside as potential threats to production levels or as additional work generating costs but not profits. The only exception was a few environment-friendly farming practices generating profits, which developed rapidly over the past few years. In Jiangxi, for instance, I could acknowledge a prompt development of the use of organic fertilizers, as it was nurturing the development of a whole new sector creating employment, generating economic growth, and attracting investors. In Jiangxi again, pest traps equipped with solar panels and manufactured by local companies or subsidiaries were often observable on orange tree plantations (Fig. 6.1).

The productivist framework promoted by central and local agricultural policies is very similar to the one spread by the Green Revolution in the aftermath of World War II. This movement, by focusing on improving the yield of crops such as maize, wheat, and rice, through technology such as improved seeds, inputs, and mechanization, contributed to the rapid spreading of monocropping. Monocropping was an ideal ally to the spreading of the technological paradigm of the Green Revolution and perfectly compatible with the objective to raise the living level of farmers. Allowing for economies of scale, monocropping also enabled specialization and investment in new technology, theoretically leading to high returns for farmers. However, a number of experts recently started ringing the alarm bell, denouncing the detrimental effects of this productivist paradigm inherited from the Green Revolution. Monocropping, in particular, has been denounced as responsible for fertility losses and land degradation under various contexts (usually resulting in higher application of nutrients and water contamination) (IPES-Food 2016) and vulnerability to biotic and abiotic stresses (Zhu et al. 2000; Jones et al. 2013) (usually in higher consumption of pesticides and herbicides, resulting in biodiversity losses), also casting doubts over whether the high productivity rates that resulted from the Green Revolution can be sustained in the future (Ray et al. 2012; Thornton 2010).

Fig. 6.1   Pest trap in orange field in Jiangxi (Photography by the author, October 2012)

The uniformization and standardization of production modes through monocropping is widely praised in China. Under "one village one product" programs (一村一品 *yicun yipin*), local officials encourage agroindustrial players to develop business models for a limited number of commodities, usually selected according to the history of the area, to its comparative advantages or to opportunist strategies. Orchards I visited in Jiangxi were exclusively producing oranges and pomelos, whereas orchards in Shandong were specialized in the production of a single variety of apples. Such governmental programs promoting the production and branding of specific regional products (of which the rationale shares similarities with the one of geographical indications, also under development in China) are in the interest of agroindustries. These latest indeed usually invest in machinery and processing equipment that are adapted to the production of a restricted number of commodities. In addition, food-processing industries have widely developed business models, which target clients who are usually willing to buy large volumes of a few products. In the other areas I visited—except from the horticultural farms around Beijing—investors were usually coming in rural areas with the idea of growing a single type of crop.

Small farmers, on the opposite, usually cultivate a wider variety of products. A lot of farmers, for instance, still grow vegetables for their own consumption. In addition, plots are usually small and scattered over a wide geographical area, because of rural–urban migration and redistribution processes occurring at the occasion of life events. The small size of plots and the fact that they are not grouped together usually encourage farmers to cultivate several varieties. In Anhui, for instance, farmers were usually cultivating rice and cotton, whereas in Chongqing, they were growing maize and pepper. On the opposite, industrial players usually look for large plots to be able to mechanize the production of a single type of commodity and achieve economies of scale. Whereas in the past, crop rotations and other agroecological practices used to be widespread throughout the whole country (King 1949) (and persist today in many areas), they are usually disregarded by industrial players.

In Jiangxi, many entrepreneurs I met had bought forest land from the government in the middle of the 2000s. As fruit orchards are classified as forests, local governments have been able to sell hilly areas covered with wild trees to investors of the fruit business without changing the official land use. While land use did not change on paper, such sales had considerable effects on local biodiversity. In addition, a vast majority of orchards are treated with herbicides and pesticides, further contributing to biodiversity losses. Such effects on the environment have extremely adverse consequences on farmers themselves. During fieldwork in Jiangxi province, farmers, agribusinessmen, and local officials sometimes desperately asked for solutions to address the yellow dragon disease, which was extremely rapidly spreading to local orchards. As no chemical existed to fight the disease at that time, the only solution was to cut down entire orchards, leaving farmers and businessmen who had invested for several years to grow trees with nothing but hills covered with stumps.

The debate on the impact of agricultural intensification and industrialized and large-scale production models on the environment is not unique to China. Defenders of productivism argue that the intensification of agriculture "spares land" through increases in yields, and, as a consequence, would in the end have positive effects on the protection of the environment (Borlaug 2002)—and therefore on biodiversity. However, this theory, which relies on a fixed demand of food, has been strongly put back into question in recent years (Pirard and Treyer 2010). A major study conducted by an interdisciplinary team of researchers in 2009 using 1990–2005 FAO data for 161 countries and 10 major crop types concluded that "agricultural intensification was not generally accompanied by decline or stasis in cropland area at a national scale" (Rudel et al. 2009: 20, 675). Angelsen and Kaimowitz (2001) go a step further by stating that increases in yields can result in the expansion of cultivated areas and environmental degradation.

A final remark is that it would be untrue to say that agricultural intensification is always linked to the transformation of small agricultural structures into industrialized and large-scale production models. In China, the tremendous rise in the use of pesticides and fertilizers started as soon as the 1970s, when small agricultural structures still constituted the vast majority of farms. Situations of agricultural production are extremely diverse across countries and one should be cautious not to generalize the conclusions drawn from the analysis of a limited number of areas. However, the example of the conversion of forests into fruit orchards in Jiangxi at the beginning of the 2000s proves that the involvement of industrial players raises questions about the environmental sustainability of the model, and the conservation of biodiversity in particular.

6.3 The Effect of Local Patterns of Power on the Sustainability of Farming Practices

Development policy is due for its own redesign based on careful consideration of human factors. World Development Report (2015).

Food-processing enterprises, retailers and, to a lesser extent, agrochemical companies are increasingly encouraged to conduct trainings in rural areas, in order to improve farming practices and the safety of food products. It is indeed in their interest to market safer products in a context where the concerns of consumers are keeping on rising. Interviews and observations on the field showed that an increasing number of trainings were indeed provided in rural areas. Numerous food-processing enterprises I met were providing trainings to farmers. In addition, they were actively trying to implement management methods including technical advice provided by their own staff. Retailers conducting direct purchase projects in rural areas were also quite active.

Direct purchase is usually not just about buying products directly from rural producers at better prices. It is also about improving the quality of products through closer management of rural suppliers. Retailers involved in direct purchase increasingly carry out trainings for farmers and factory managers. This is particularly true for foreign retailers, which are more exposed to bad publicity in the Chinese media compared to local supermarkets.[1] As a consequence, foreign retailers are particularly eager to invest efforts in improving their image and relationship with officials, for whom taking an active part in agricultural modernization can be really helpful.

Trainings targeting factory managers usually aim at helping them make food processing better equipped to address the demand of retailers in terms of volumes, traceability, and safety. Trainings linked to farming practices are supposed to push farmers to adopt more sustainable practices and to implement traceability methods. From 2007 to 2013, more than 50 trainings were conducted by retailers, mostly in the framework of direct purchase projects (Hu 2013).[2] The ones I could attend to, even though I was told participants were farmers, were in reality mostly gathering factory managers and managers of farm associations. In the case of trainings linked to DP projects, managers of factories and farm associations in fact often act as transmission belts between experts mobilized by retailers and farmers working on the field.

Trainings linked to direct purchase are sometimes deeply intertwined with the contracts established between retailers and rural factories, which include development plans. The trainers that are mobilized by retailers include members of retailers' quality teams as well as Chinese scientists (regional experts and renowned researchers from Chinese universities) and focus on environmental as well as social aspects, as this quote of a trainer illustrates:

> The first day of the training, I explained the requirements in terms of production practice: herbicides are forbidden [...], they should not use hormones either, more than 60 percent of fertilizers have to be organic... We also have social requirements. (Interview with trainer belonging to X. retailer's quality team, Shanghai, October 2013)

Trainings are not always enthusiastically welcomed by industrial managers, who fear to invest in the modernization of their practices and processes without being adequately compensated. They perceive trainings as additional time- and money-consuming requirements made by clients who then refuse to pay more for upgraded products. Retailers, on their side, provide trainings for free to industrial players, with the idea that they are helping them modernize their process and manufacture products for the mass-market retailing, not with the aim of creating a niche market of luxury products for wealthy consumers. The fact that long-term contracts are often associated with trainings does not reassure industrial players, who remain suspicious about the long-term engagement of retailers. A quote from a manager of X. in charge of conducting trainings in rural food processing enterprises sums up these differences in points of view:

> For the first training, I just talked about production processes. If we start talking about traceability, they have dollars in their eyes because for them, traceability equals high-end products [more expensive]. [...] [During the first training, I told them that] more than 60 percent of fertilizers had to be organic. They listened to us, and after that they all said that they were using exactly 60 percent of organic fertilizer in their fields! [...] We spend time explaining the whole philosophy of the project to them. We try to make them understand what will be the benefits for them: that they will have an edge over other suppliers, that they will be able to export, etc. (Interview, Shanghai, October 2013)

Although industrial players and retailers share different views, their involvement in agricultural modernization could still have positive effects on farming practices through the multiplication of trainings. As it is in the interest of industrial players and retailers to sell safer products and as these latest have the financial capacity to recruit skilled trainers to disseminate knowledge, farming practices should theoretically evolve toward more social and environmental sustainability. However, the fact that farmers remain marginalized in the process of agricultural modernization considerably lowers the possibility of a real evolution of farming practices, for three main reasons.

The first reason is that the increased involvement of food-processing enterprises in agricultural modernization—through the subrenting of farmland, the contractualization with small farmers, or the increased trainings on farming practices—does not change the set of interests of farmers, and, as a consequence, does not encourage them to change their practices. In Lushan and Lanshui, farmers are usually still paid according to the weight and to the quality—mostly referring to the appearance of fruits—of products they are able to yield for the factory, even when they are shareholders in a farmers' cooperative that is associated with the factory. As a consequence, farmers are reluctant to use less pesticides and fertilizers, as they face the risk to decrease their yield or to affect the appearance of their fruits and, as a consequence, to be paid less. It is not a risk they are willing to take, considering their already low income and the absence of insurance coverage.

The second reason is linked to the fact that, in the areas where I conducted fieldwork, trainers were keeping on relying on traditional top-down approaches of teaching. Such practices are already widely used by local officials. In Chongqing, for instance, the setting up of hotlines by agricultural extension services bureaus was supposed to link farmers with technical experts, but in reality only reinforced the distance existing between experts and farmers. In Anhui, local officials dedicate important efforts to prevent farmers from burning straw in order to limit greenhouse gases emissions. However, in a village near Hefei (in Anhui province), the methods used by the government—a car passing by villages or parked in front of farmers' markets with a loudspeaker repeating to "friend-peasants" not to burn straw (Fig. 6.2)—let me rather puzzled. The distance put between officials and trainers on one side, and farmers on the other side, constitute a considerable obstacle impeding the coop-

Fig. 6.2 Governmental car with a loudspeaker parked in front of the farmers' market

eration of farmers and the evolution of farming practices. The effectiveness of such top-down methods for transition is highly questionable. In Chongqing, farmers were not calling hotlines. In Anhui, during the evening meals, gathered villagers were devising ways to burn straw that would prevent officials from noticing and were wondering about how to disturb the car during its next visit.

International organizations, development agencies, and an increasing number of countries have adopted the rhetoric of participatory development as a way to achieve greater sustainability of projects and to efficiently steer transition. The 1992 Rio Declaration on Environment and Development, for instance, states that "environmental issues are best handled with the participation of all concerned citizens, at the relevant level. At the national level, each individual shall have […] the opportunity to participate in decision-making processes." Participatory processes have multiplied worldwide, particularly in the field of environment and sustainable development (Hamdouch and Zuindeau 2010). While a large corpus of literature questions the effectiveness of participatory approach for transition, there seem to be at least a consensus about the low efficiency of top-down approach in the field of transition toward more sustainability. For Van Tatenhove and Leroy (2003), participation is inextricably linked to environmental issues. However, in China, rural enterprises that were investigated were keeping on using traditional top-down methods for the spreading of agricultural knowledge and insights from fieldwork showed that such an approach was quite inefficient at changing practices. In reality, such methods only increase the rigidity of the barriers that exist between the different social layers. Factories are increasingly trying to implement close-management methods. Some hire technicians in charge of managing small groups of farmers in fields. Others purchase pesticides and fertilizers directly for farmers. Sometimes, the "best" farmer is awarded a position of technical management. However, in most of the situations I encountered, exchanges between circles of stakeholders—the one of *nongmin* growing products in the field and the one of managers or entrepreneurs from "upper-levels"—remained poor: in one way, as it was very difficult for a *nongmin* to become a manager or an agricultural businessman; *and* in the other way, as instructions coming from upper circles are disregarded by farmers. Misunderstanding and lack of efforts to listen to the other groups are frequent, as this amazing anecdote illustrates:

> This year, they [(the factory managers)] put a manager in charge of every district. In each district, the manager does "close management": they do a lot of meetings, so that they can teach the peasants how to use this pesticide or that. But […] farmers […] did not tell me the same thing as [the company] did, they told me that they had always grew trees on their own, that they knew how to plant trees and did not need any advice from them. Today, they launch some meetings, but maybe some farmers will listen to them, maybe some others won't. Last season, they tried to improve the results of farmers by launching a contest: the best farmers would go to Xiamen free of charge, but farmers did not understand. They thought that maybe they would have to pay something, so they did not want to participate to the best performance contest, but then they learnt about the "free of charge", and they regretted it. (Interview with quality auditor, Jiangxi, October 2012)

The last reason that explains why the marginalization of farmers has a strong effect on their unwillingness to change their farming practices is that *nongmin* are locked up in an isolated circle way too far from consumers. As food-processing enterprises have become nonremovable intermediaries of the food chain, small farmers, remote in rural areas, can easily hide behind these factories, thus contributing to isolate them even more from the consumers and from these latest' concerns in terms of food safety.

The most recent surveys evaluating pesticides residue in fruits and vegetables confirm that major problems still exist on the side of farming practices, which remain detrimental both to the environment and to the health of consumers. The results of an investigation on pesticides residue in fruits and vegetables conducted by the AQSIQ in 2014 in 23 major Chinese cities are alarming, with highest passing rate at 72.4 percent and lowest passing rates at 47.5 percent (Hao 2014).

6.4   WHO WILL FARM IN THE FUTURE?

Persistent inequalities between rural and urban areas are reinforcing the strong distaste of young and active people for farming, and more generally for living in rural areas. The Chinese government wishes to encourage rural–urban migrations, to a certain extent, in order to increase the size of farming structures, so that they could be modernized, mechanized, create economies of scale, and help farmers raise their income. However, the development of "professional and modern farms", highly desired by the government, requires the involvement of an active and educated labor force. The problem is that the poor living conditions in the countryside do not encourage young and educated people to launch businesses in rural areas, especially in the farming sector, which is particularly despised. Although the relative gap between urban and rural revenue decreased over the recent years (the ratio went from 3.23 in 2010 to 3.1 in 2012), in terms of absolute value, the gap actually kept on growing (going from 131,901 RMB in 2010 to 16,648 RMB on the same year). In addition, official data released by the National Bureau of Statistics probably underestimate the real gap. The calculation of revenue, in rural areas, sometimes includes self-consumption, such as grain and vegetables grown by the farmers themselves and consumed by the household members. On the opposite, revenues of urban dwellers do not include subventions they get from unemployment and health insurance. In 2012, the net income per capita, in rural areas, was of 7917 RMB per year (by comparison, the disposable income per capita in urban areas was of 24,765 RMB).

Although agricultural universities exist (there are no less than 20,000 students at the Chinese Agricultural University in Beijing), the attractiveness of the city life—both in terms of revenue and in terms of living conditions—encourage students majoring in agronomy to look for jobs in research institutes in biotechnology, in political bureaus or in sectors not related to agriculture at all. As was saying a young woman graduated from the Agricultural University of Hebei province, now working in the purchase department of retail company in Shanghai:

> My major was agriculture. Most of my former schoolmates now work in Beijing, but in fields completely other than agriculture. (Interview, Shanghai, October 2012)

Another former student having majored in agronomy, graduated from the Chinese Agricultural University, had decided to turn to marketing after unsuccessful experiences in the agricultural sector (Interview, Beijing, May 2013). According to a Master's director at the Chinese Agricultural University, the situation is just like what his students had depicted, but is currently evolving:

> All of my students have found a job. There are more and more jobs for them, because there are more and more enterprises linked to agriculture. (Interview, Beijing, May 2013)

In order to encourage young people to live in rural areas, the government has been actively trying to improve rural living conditions through the development of infrastructures and the rise in agricultural subsidies. A number of programs aimed at training on-site farmers have also developed over the past few years. In 2004, the government created the "Sunshine project" (阳光工程 *yangguang gongcheng*). At first, this program aimed at providing rural dwellers with trainings linked to catering and hotel services, health care, construction, manufacturing, and domestic service, in order to lift them out of poverty by offering them the necessary background to work in sectors other than farming (People Daily 2004). In March 2013, the Sunshine project was revised, giving a much larger role to the agricultural sector. The statement published by the MOA in 2013 clearly reflects this shift in priorities: "The 'Sunshine Project', a project designed to train rural labors for increasing their employment opportunities in cities, is to be reoriented to training on agricultural technology and agribusiness." In addition to the Sunshine Project, several programs seek to educate and train on-site and new farmers and at "enhancing the development of rural talents". Among others, programs include business start-up trainings, basic scientific education, and field visits.

Efforts to raise the attractiveness of the farming sector and to improve the knowledge of on-site farmers are important. However, fixing people in rural areas and ensuring that these latest are both able to take care of the land that is abandoned by migrants and active enough to modernize the sector is a rather challenging task. Even if economic policies would be perfectly efficient in filling infrastructure and income gaps between rural and urban areas, it is not sure that this would be enough to get rid of the cultural factors stigmatizing rural areas and the farming sector in the mind of the population.

The always-greater flow of farming workforce escaping from rural areas raises the question of "who will farm in the future". On a more shortcoming perspective, the fact that rural areas are highly unattractive to young and active people raises the question of "who will be able to modernize the agricultural sector". The marginalization of farmers in the modernization process leads to the impossibility for *nongmin* to access better social and economic conditions through farming and encourages them to adopt a going-out strategy. By escaping the farming sector and rural areas, they have a greater chance of being freed from their social condition and to have access to better living conditions. Insights from fieldwork showed that one of the main consequences of the fact that farmers were privileging a going-out strategy and that local officials were quite eager to see them leaving the farming sector, was that it was increasingly difficult for agri-food enterprises to find labor force for farming in a number of places. Most of the agricultural workers I met were about 50 or 60 years old (Figs. 6.3, 6.4, and 6.5). For these farmers, looking for jobs in cities is barely an option, considering their age and health condition—and sometimes, their engagement to look after their grandchildren in the countryside. It is very unlikely that their children will come back to the farming sector in the future, considering the conditions the job of farmer-worker currently offers them: seasonal work, low income without health coverage and retirement pension, and, above all, the absence of opportunities for career development and for the improvement of their living conditions—which is among the most important incentives for young rural dwellers to become migrants. As

Fig. 6.3 Farmers-workers wrapping oranges in plastic bags in a factory, Jiangxi (Photography by the author, Oct. 2013)

a consequence, questions arise about the sustainability of the farming workforce, and, as a corollary, about the sustainability of agricultural production in the future. With or without the establishment of land ownership, the emergence of modern farmers able to take part in the process and become real levers of agricultural modernization will be one of the most important and arduous challenges of agricultural modernization in China.

The urban–rural dual structure has increasingly been denounced by the Chinese media and by a number of scholars (Lan 2013; Fan 2012; Lu 2012), as an unfair system likely to trigger social instability and to impede economic growth. Since 2008, the central government, aware of the situation, has been regularly trying to encourage local governments to make their *hukou* scheme more flexible and to integrate more migrants in their urban systems. However, local governments are still reluctant to give up on the *hukou* system. According to a report issued by the China International Research Committee for Development and Urbanization Strategy (2008), the average cost of integrating one rural migrant into urban systems would be 100,000 RMB. In addition to the expenditures linked to the integration of migrants into education and health systems, costs linked to the building of new infrastructures—such as roads or electricity and water networks—should also be added. Local governments

Fig. 6.4 Farmers collecting oranges on the land of a factory in Jiangxi (Photography by the author, Oct. 2013)

Fig. 6.5 Farmer collecting oranges on the land of a factory in Jiangxi (Photography by the author, Oct. 2013)

would also be afraid that the integration of rural migrants into urban systems would lower the quality of existing public services and provoke anger among the connected urban middle- and upper-classes, who have the means to protest against the downgrading of public services. Finally, the urban–rural dual structure created by the *hukou* system has also long been granting enterprises established in urban areas with an abundant and cheap labor force coming from rural areas, hard to give up onto. Debates have been fiercer lately about the relaxation of the *hukou* system. Although this is an unquestionable proof of the willingness of the central government to improve the living conditions of the population of migrants, the reluctance of local governments to change the system is still strong, particularly in the overpopulated cities of the eastern parts of the country, where the majority of the urban population is concentrated.

In the past, the floating population used to enjoy the possibility to go back to the countryside to farm its land and have access to subsistence agriculture, in case of dismissal or disease. However, the new generation of migrants has little experience in farming and has higher expectancies of urban life (Zhang 2013). In other words, most of them do not intend to return to the countryside, even on a temporary basis.

In cities, migrants live in rented insecure accommodation, sometimes illegal (e.g., 地下室 *dixiashi*, or underground housing), sometimes not connected to basic public infrastructures. Local authorities often see these places as slums (贫民窟 *pinminku*) that need to be turned down for the sake of modernization of urban areas. As stated by Wu et al. (2014: 13): "The habitats of rural migrants are still regarded as backward places to be modernized." Although programs aimed at compensating and relocating people expelled from their houses do exist, they usually do not include migrants in their beneficiaries.

The inequalities between these two groups of urban dwellers who live side-by-side (or beneath one another) are likely to give rise to social unrest. Violent protests already regularly occur on building sites, where workers express their anger for delays in the payment of their salaries. Today, another issue is increasingly worrying local authorities. The average "migrant household" size has increased to 2.5 persons, as less and less migrants leave their children to their parents staying in rural areas. This new generation of young migrants has never farmed, sometimes never lived in the countryside, and compares its living conditions with the ones of other urban dwellers. This might create a feeling a frustration that can lead to important social unrest. As Pun and Lun (2010: 512) sum it up:

The second generation of peasant-workers has gradually become aware of its class position and has participated in a series of collective actions. Having a quasi-social status, *nongmingong*, the second generation of migrant workers is now experiencing a deeper sense of anger and dissatisfaction than that of the first generation, and is realizing that they are increasingly cut off from so many erstwhile or nominal sources of support—in fact, there is almost no returning to their hometown.

Whereas in the past, the conditions of migrant workers contributed to the Chinese economic development by offering cheap labor to industries and urban construction sites (migrants were accounting for 68 percent of the employees of processing and manufacturing industries and 80 percent of the labor of the construction industry (Chinese State Council 2006: 172)), they are now threatening social stability in urban areas and preventing the country from benefiting from a leverage effect for economic growth. The floating population indeed accounts for almost 20 percent of the Chinese population. Migrants do not enjoy high salaries and have to spare money in case of dismissal, disease, or retirement. As a consequence, they consume much less than other urban dwellers, although having access to the same markets and commodities. In the current context of contracted international demand, the Chinese government is perfectly aware of the necessity to rely on domestic sources to generate economic growth. Since the middle of the 2000s, several regulations and policies were issued that aimed at improving the living conditions of the 230 million of potential consumers living in cities (Zhang 2013: 170). Stabilizing the living conditions of the population of migrants is essential, not only to address social issues in urban areas but also because it would encourage migrants to give up on their land, enabling farmers staying in the countryside to cultivate bigger farms.

Rapid rural–urban migrations that are emptying a lot of rural areas from their labor force without necessarily bringing additional land to the market (that could enable farmers staying in the countryside to cultivate bigger farms) have adverse social effects that go well beyond rural areas. The question indeed arises, whether a slowing down economy will be able to provide jobs for the hundreds of millions of people that local officials are pushing out of the farming sector on a permanent basis, so that they would give up on their land. Until today, the booming industrial sector has provided farming labor surplus with jobs that did not require specific skills. However, questions about the sustainability of the Chinese model of growth arise today. First, the worldwide economic slowdown has already had negative effects on the country's exports—even though the recent recovery of the US economy holds out the hope that exports recover for a longer period of time. In addition, the growing environmental concerns that are linked to industrial development question the entire model of the Chinese economy. For a number of scholars indeed, the slowdown of industrial growth resulting from aggravating environmental issues is likely to become a worldwide trend in the next decades. As Dorin et al. (2013: 16) put it: "Industrial production might increase more slowly in the future due to the increasing cost of oil and other non-renewable resources, strengthened environmental regulations, market saturation in industrialized countries, and slower wage increases in developed economies not fully compensated for by an increase of incomes in developing countries."

The dominant frame of reference of agricultural modernization, in China, strongly echoes the theory of the Lewis path for development. According to the Lewis path, agriculture is the first step of economic development, as agriculture provides labor, savings and low-cost food to the process of industrialization and urbanization. Industry, in turn, is supposed to provide increasingly cheaper agricultural inputs that improve yields, rising labor productivity of the rural economy, and, consequently, drawing up wages and eliminating poverty. As summed up by Dorin et al. (2013: 3), the Lewis path is "anchored in economic theories about interrelated structural changes between the 'traditional' (agriculture) and 'modern' (non-agriculture) sectors and in the historical experience of 'modern economic growth'" and ultimately leads to "world without agriculture"—a theory of modernity which, by the way, shares interesting similarities with the concept of *suzhi*. In their analysis, the three researchers argue that the Lewis path is actually one path among four contrasting developmental paths that do not necessarily converge. The authors demonstrate that switching from one path to a Lewis path can be difficult—if not impossible—for a number of countries. For instance, they argue that mega-urbanization will lead to considerable challenges in emerging countries. While the Lewis Path was facilitated in European countries, where cities managed to retain low-density populations thanks to the migration of 60 million people to the "New Worlds", this possibility is not offered to developing and emerging countries, where urban space is continuously shrinking. Drawing on the example of India, the authors build up two alternative scenarios. In the first variant, that they call "Lewis trap," farmers cannot migrate rapidly enough to crowded urban shantytowns and are "condemned to stay with a business whose natural capital declines (soil, biodiversity, safe water) while their own capabilities are diminished due to poverty (nutrition, health, education)" (Dorin et al. 2013: 16). The resulting growth of disparity between rural and urban areas then puts high-performing Asian economies at the risk of facing severe social crises coming from the countryside, likely to threaten their economic development. In the second variant, the disparity problem transfers to cities, with the coexistence, in urban areas, of highly skilled and highly paid labor with highly labor-intensive and low-wage services, leading to similar social issues. For the authors (Dorin et al. 2013: 18), "Asia cannot replicate [the] experience [of Western countries] nor share the utopia of a few large-scale farmers and agro-industries feeding the bulk of humankind in huge megacities". Although the authors mainly build their analysis by drawing on the example of India, their conclusions match the conclusions drawn by this research on the analysis of Chinese case studies. If China persists in trying to apply the Western model of agricultural modernization on its territory, the country is likely to let two development pathways emerge, both socially and economically unsustainable: the first variant of the "Lewis trap", where farmers, unable to find jobs outside the farming sector, would be condemned to stay in rural areas, where they would not be able to make a living out of farming; the second variant of the Lewis trap, where migrants would leave the countryside (raising issues in terms of agricultural labor) and would come to live in cities where they would coexist with well-off citizens with whom they would not share the same rights. In a vast country such as China, it is probable that the two variants emerge at the same time, in different areas of the territory.

Similar conclusions about the impossibility of the Western model for agricultural modernization to fit China's specificities were also reached by a number of Chinese scholars, as the following sentence, which quotes a scholar of the CASS, illustrates (Yan and Chen 2013: 966):

> What is becoming shared knowledge among many rural support intellectuals is clearly stated by Yang Tuan, a scholar at CASS. The US/Western model, associated with de-peasantization [*qu nongmin hua*], industrialization, and urbanization, is a model that works for a small number of capitalist farmers and corporations who enjoy big government subsidies, while China needs to find a way to sustain a large rural population.

In spite of the clear limits the lever of rural exodus has in the process of agricultural modernization, the belief in the efficiency of the lever remains strong among local officials.

6.5   CONCLUSION

In spite of an increasing willingness of the government to address issues such as environmental protection and food safety—which were mentioned in central documents as soon as in the beginning of the 2000s—central guidelines referring to these issues do not have the importance of the ones urging for the improvement of food security or rural living standards. The efficient spreading of certain elements of the frame of reference designed by the central government has consequences on the agricultural modernization pathway, in the sense that it prevents the country from engaging on a path toward environmentally and socially more sustainable farming practices. The great emphasis put on the goal of food security and the strategy to rely preferentially on industrial players to trigger modernization in rural areas have particularly strong effects. During my fieldwork, I could acknowledge that food-processing enterprises and retailers were encouraged to conduct trainings to improve farming practices and the safety of food products. Interviews and fieldwork showed that an increasing number of trainings were indeed provided in rural areas. However, insights from fieldwork also demonstrated that the marginalization of farmers in the process of agricultural modernization considerably lowered the possibility of a real evolution of farming practices. The socio-economic models of agricultural modernization indeed do not change the set of interests of farmers-workers, who are usually still paid according to the weight and appearance of the products they harvest. In addition, rural enterprises which conduct trainings in rural areas usually keep on relying on top-down methods for the spreading of agricultural knowledge, which maintain farmers-workers in a low social class that is strongly bounded, difficult to escape from and also difficult to reach from above. Farmers, isolated in their condition, are way too far from the concerns of consumers in terms of food safety to start thinking about changing their practices. Finally, the overreliance on industrialized and large-scale agri-food enterprises looking for economies of scale through monocropping is likely to have adverse effects on biodiversity—a situation which is not unique to China.

The marginalization of farmers and the rapid spreading of the dominant frame of reference for agricultural modernization are also likely to have effect on the sustainability of agricultural output in the future. Indeed, the strong marginalization of farmers does not offer them the possibility to escape from their low social and economic condition through farming, encouraging them to adopt a going-out strategy. In addition, the spreading of the idea that modern agriculture necessitates the migration of rural dwellers to cities in order to enable farmers staying in the countryside to cultivate wider areas of land is likely to further deprive the agricultural sector from a precious labor force of young and educated rural dwellers, which raises questions about the sustainability of food production in the middle and long term. Finally, social challenges are also likely to arise. In case the slowing down economy turns unable to provide jobs on a permanent basis for the hundreds of millions of people local officials wish to force out of the farming sector, consequences are likely to be severe in rural areas. In case farmers effectively migrate to cities without having access to the same rights of their urban neighbors, social consequences will probably spread beyond rural areas.

NOTES

1. Some foreign supermarkets are really seen as the flagships of their country of origin, such as demonstrated the 2008s boycott of Carrefour's products and protests in front of its supermarkets, in response to French pro-Tibet demonstrations during the summer Olympic torch relay.
2. Other data were collected during fieldwork showed that X. conducted 12 trainings in 2011 and 14 trainings in 2012 and that another foreign supermarket was conducting approximately the same number of trainings (the two supermarkets were though probably the most active ones in the field of farmers' trainings).

REFERENCES

Angelsen, A., & Kaimowitz, D. (Eds.). (2001). *Agricultural technologies and tropical deforestation*. Wallingford: CAB International.

Borlaug, N. (2002). Feeding a world of 10 billion people: The miracle ahead. *Vitro Cellular & Developmental Biology—Plant, 38*(2), 221–228.

Chinese State Council (Research Office). (2006). *Research Report on Migrant Workers*. 2006 [国务院研究课题室, 中国农民工调研报告, 北京: 言实出版社, 2006 *Guowuyuan yanjiu keti shi, zhongguo nongmingong diaoyan baogao, beijing : yanshi chubanshe*].

Dorin, B., Hourcade, J. C., & Benoit-Cattin, M. (2013). *A world without farmers? The Lewis path revisited*. CIRED working papers 47.

Fan, Z. (2012, July 19, 29). Accelerate the transformation of rural migrants into urban residents. Caijing [范子英, 加速农民工"市民化", 财经, 2012年第19期 *Fan Ziying, jiasu nongmingong 'shiminhua', Caijing*].

Hamdouch, A., & Zuindeau, B. (2010). Introduction. Diversité territoriale et dynamiques socio-institutionnelles du développement durable : une mise en perspective. *Géographie, économie, société, 12*(3), 243–259.

Hao, X. (2014, June 20). Pesticide-degrading enzyme to improve food safety. *Journal of Science and Technology* [郝晓明, 农药解毒酶让"舌尖"更安全, 科技日报, 2014年06月20日 *Hao Xiaoming, Nongyao jiedu mei rang "shejian" geng anquan, Keji ribao, 2014/06/20*]. http://digitalpaper.stdaily.com/http_www.kjrb.com/kjrb/html/2014-06/20/content_266279.htm?div=-1. The article also appears on the website of the People's Journal. http://scitech.people.com.cn/n/2014/0620/c1057-25177468.html

Hu, D. (2013, November 13). *The opportunity & challenges of farmer-supermarket direct purchase in China*. Paper presented at the FAO's Policy Forum on Rural–Urban Income Gaps and Smallholder Market Integration in Asia, Beijing.

IPES-Food. (2016). *From uniformity to diversity: A paradigm shift from industrial agriculture to diversified agroecological systems*. International Panel of Experts on Sustainable Food systems.

Jones, B. A., Grace, D., Kock, R., Alonso, S., Rushton, J., Said, M. Y., McKeever, D., Mutua, F., Young, J., Mc-Dermott, J., & Pfeiffer, D. U. (2013). Zoonosis emergence linked to agricultural intensification and environmental change. *Proceedings of the National Academy of Sciences of the United States of America, 110*, 8399–8404. https://doi.org/10.1073/pnas.1208059110.

King, F. (1949). *Farmers of forty centuries; or, permanent agriculture in China, Korea and Japan*. London: J. Cape.

Lan, F. (2013, January 1, 7). The gap between urban and rural areas: The barriers to the dismantlement of transformation of rural migrants into urban residents. Caixin New Century [蓝方, 城乡鸿沟:农民市民化中那些待拆的壁垒, 财新《新世纪》, 07/01/2013 *Lan Fang, chengxiang honggou: nongmin shimin hua zhong naxie dai chai de bilei, Cai xin "Xin shiji"*].

Lu, M. (2012, July 16, 18). The economic benefits of the freedom of movement. Caijing [陆铭, 自由迁徙的经济价值, 财经, 2012年第18期 *Lu Ming, Ziyou qianxi de jingji jiazhi, Caijing*].

People Daily. (2004, February 18). 'Sunshine' project to offer rural people job training. *People Daily*. http://english.people.com.cn/200402/18/eng20040218_135081.shtml

Pirard, R., & Treyer, S. (2010). Agriculture et déforestation: quel rôle pour REDD+ et les politiques publiques d'accompagnement? *Iddri—Idées pour le débat, 10*, 1–18.

Pun, N., & Lun, H. (2010). Unfinished Proletarianization: Self, anger, and class action among the second generation of peasant-workers in present-day China. *Modern China,36*(5),493–519.https://doi.org/10.1177/0097700410373576.

Ray, D. K., Ramankutty, N., Mueller, N. D., West, P. C., & Foley, J. A. (2012). Recent patterns of crop yield growth and stagnation. *Nature Communications, 3*. https://doi.org/10.1038/ncomms2296.

Rudel, T. K., Schneider, L., Uriarte, M., Turner, B. L., Defriesc, R., Lawrence, D., Goeghegan, J., Hecht, S., Ickowitz, A., Lambinh, E. F., Birkenholtz, T., Baptistai, S., & Grauj, R. (2009). Agricultural intensification and changes in cultivated areas, 1970–2005. *Proceedings of the National Academy of Sciences, 106*(49), 20675–20680.

Thornton, P. M. (2010). Livestock production: Recent trends, future prospects. *Philosophical Transactions of the Royal Society of London B: Biological Sciences, 365*, 2853–2867. https://doi.org/10.1098/rstb.2010.0134.

Van Tatenhove, J. P. M., & Leroy, P. (2003). Environment and participation in a context of political modernisation. *Environmental Values, 12*(2), 155–174.

World Bank. (2015). *World development report: Mind, society and behavior.* Washington, DC: World Bank.

Wu, F., Zhang, F., & Webster, C. (Eds.). (2014). *Rural migrants in urban China: Enclaves and transient urbanism.* London/New York: Routledge.

Yan, H., & Chen, Y. (2013). Debating the rural cooperative movement in China, the past and the present. *Journal of Peasant Studies, 40*(6), 955–981. https://doi.org/10.1080/03066150.2013.866555.

Zhang, X. (2013). The new generation of migrant workers in labour market in China. In L. Pries (Ed.), *Shifting boundaries of belonging and new migration dynamics in Europe and China.* Basingstoke: Palgrave Macmillan.

Zhu, Y., Chen, H., Fan, J., Wang, Y., Li, Y., Chen, J., Fan, J., Yang, S., Hu, L., Leung, H., Mew, T. W., Teng, P. S., Wang, Z., & Mundt, C. C. (2000). Genetic diversity and disease control in rice. *Nature, 406*, 718–722. https://doi.org/10.1038/35021046.

CHAPTER 7

Conclusion

7.1 A Dominant Frame of Reference

In spite of the fundamental importance rural areas, agriculture and peasants had in the building of the CCP, the Chinese countryside had cruelly lost the interest of the government in the late twentieth century. Local cadres had shifted their attention to industrial development as a way to steer economic growth in rural areas and to keep the political and economic control they had in the era of People's Communes. The central government, on its side, pressured by the growing stakes of industrial development and urbanization, had also turned its focus to such sectors and lost interest in rural areas and agricultural development.

At the beginning of the twenty-first century however, faced with rising challenges in terms of food security, social stability, and economic development, the central institutions of the government started encouraging local officials to reinvest efforts in agricultural production, a sector they had deserted since the household responsibility system had put an end to planning and given back the reign of agricultural production choices to farmers at the beginning of the 1980s. In particular, from 2004 onward, central institutions started promoting policy guidelines urging local officials to speed up agricultural modernization. These directives, promulgated, among others, through Five-Year Plans and Number One Documents, progressively built a frame of reference for

© The Author(s) 2018   237
M.-H. Schwoob, *Food Security and the Modernisation Pathway in China*, Critical Studies of the Asia-Pacific,
https://doi.org/10.1007/978-3-319-65702-8_7

agricultural modernization, defined by: (i) two key goals: food security (in the sense of self-sufficiency) and rural development; and (ii) three key levers: scientific and technological development, enterprises (and especially dragonhead enterprises) and the rural exodus.

Local officials followed the directives of the central government pushing them to reestablish links with the agricultural sector and to steer its modernization. The reinvestment of agricultural production activities by local governments, however, was not a direct stepping, but was mostly accomplished through a strong reliance on a network of private entrepreneurs encouraged to launch agricultural business. Incentives and control mechanisms were carefully established by local officials to push and pull entrepreneurs, through the use of financial and nonfinancial resources (either material, human, reputational, or normative) both in formal ways (through standardized and institutionalized procedures) and in informal ways (where social ties are of strong importance). A form of governance mixing elements of the local developmental state, elements of the corporatist state and elements of the regulatory state emerged in the course of agricultural modernization. In rural areas, government agencies of the county and township levels act as local developmental states by selecting strategic sectors and entrepreneurs able to lead agricultural modernization. Entrepreneurs, on their side, engage in the field of opportunities offered by local governments, participate in the building of private food chains and increasingly act as trainers for farmers, thus becoming multipliers—or corporatist structures—spreading the central state's concerns down to the multiplicity of small farmers. Local officials manage to keep control over this development of private entrepreneurship by relying on regulations, like a regulatory state, but in both formal and informal ways. The possibility to decide how to apply rules is enabled by the important decentralization of the Chinese state.

While their political participation remains limited, private entrepreneurs play a major role in agricultural modernization through the launching of economic activities in rural areas and through their increasingly direct involvement in farming methods—as evidenced, for instance, by the rising number of trainings that they provide to farmers. As a consequence, we are not looking at a state-socialist economy characterized by planning anymore. Certainly, agricultural and food enterprises are firmly controlled by local governments through the use of mechanisms of which some are *legacies* of state socialism—such as the monopoly of control over political institutions (Landry 2008: 18), which, in turn, exercise power over resources. However, this research demonstrates that local officials also use a multiplicity of indirect and less visible control channels that progressively developed into close-knit communities of political and economic actors, where both formal and informal rules apply. As we see, China's agricultural modernization is a complex process that does not fit in any of the theoretical frameworks previously developed by political science—such as planned economy, developmental state, corporatist state or regulatory state. Rather, the process fits in a model made of a number of elements coming from different frameworks and helping to understand the peculiarities of agricultural modernization.

Recently, the mode of operation of local governments for agricultural modernization evolved toward a wider and more complex network of economic and political actors, which increasingly includes stakeholders downstream or upstream in the food chain, a number of which operate from urban areas. However, this does not really put back into question the assumptions mentioned above. Simply, local states become less "local" and more "transversal" and include a wider variety of actors not necessarily sharing the same interests, but still agreeing on the main principles of the mode of operation for agricultural modernization framed by the central state.

7.2   IMPACTS ON THE SUSTAINABILITY OF AGRICULTURE

The reasons for the effective transmission of the dominant frame of reference for agricultural modernization down to local levels of the government do not only lie in the efficiency of traditional transmission mechanisms. Transmission also stems from the fact that the key elements of the frame of reference match path dependencies and the interests of the local stakeholders holding power. "Classical" transmission mechanisms such as the cadre promotion system and competition between government officials do play a role in the diffusion of policy guidelines down to local levels. However, in the case of agricultural modernization, policy guidelines are efficiently spread down to local levels of the government mainly because they match local path dependencies and the interests of the most powerful stakeholders at the local level. The fact that a number of guidelines linked to environmental protection and to the development of grassroots structures are not efficiently transmitted is a clear demonstration of this assumption.

In particular, path dependency and local patterns of power play a huge role in the transmission of the element of the frame of reference granting

enterprises with a leading role in agricultural modernization. Local officials, at the beginning of the 2000s, were indeed already used to rely on
enterprises to achieve development objectives, such as the modalities of
the multiplication of TVEs in the 1980s and 1990s demonstrate. On the
opposite, they have few contacts with farmers who they usually consider as
poorly educated and refractory to modernization. In addition, relying on
food processing enterprises is in the interest of local officials because it
provides them with additional revenue through industrial and commercial
taxes, whereas farming does not since agricultural taxes were abolished in
2006.

The recent attempt of the central government to promote the establishment of farmers-led agricultural cooperatives again proves this point.
Instead of showing a mushrooming of cooperatives created by empowered
farmers, the fieldwork of this research rather demonstrates that food factories established in rural areas are often behind the development of mixed
forms of cooperatives gathering both industrial shareholders and farmers.
Inside these "farmers' cooperatives"—a number of scholars name "fake"
cooperatives—patterns of power are in fact very similar to what is usually
observed in more classical forms of association between rural food enterprises and farmers, in the sense that agricultural cooperatives maintain the
strong divide between *nongmin* and entrepreneurs/managers.

Because of a prioritization of the stake of food security in the mind
of central officials, central policies have put emphasis on agricultural
productivity over the past decades, leading to a strengthening of stakeholders (food processing enterprises in particular) who were identified
as effective "transmission belts" to rapidly pass on productive practices
to small farmers. The recent shift in central priorities, that now wish to
include more environmental considerations, proves ineffective because
of the path dependencies linked to empowered stakeholders—who
though do not have the means to make farmers shift to more sustainable farming practices, even when they conduct trainings to improve
farming practices and the safety of food products. Farmers usually
remain paid according to the weight and appearance of the products
they harvest, lowering the incentive for them to change their practices,
for instance by applying less pesticides or less fertilizer. In addition,
rural enterprises usually keep on relying on top-down methods for the
spreading of agricultural knowledge, which maintain farmers-workers
in a low social class that is strongly bounded, difficult to escape from
and also difficult to reach from above. Farmers, being marginalized in

the whole process, remain isolated from the concerns of consumers in terms of food safety, which brings even lower their incentive to change their farming practices.

Finally, the marginalization of farmers and the fact that they do not have any possibility to escape from their social and economic condition through farming encourage them to adopt a going-out strategy. They are generally encouraged to do so by local officials, who have to reach urbanization targets that go along with the idea that rural–urban migration is a necessary step for agricultural modernization, as it theoretically enables farmers staying in the countryside to cultivate wider areas of land. However, in a number of areas, this strategy has already deprived the agricultural sector from a precious labor force of young and educated rural dwellers, which raises questions about the sustainability of food production in the middle and long term. In addition, the question arises, whether the slowing down economy will be able to provide jobs for the hundreds of millions of rural dwellers willing to escape from the countryside.

7.3   Pockets of Innovation and Desertification

In spite of the spreading of a dominant frame of reference for agricultural modernization and of a common mode of operation, the analysis of cases such as agricultural development projects in Ningxia, the grain sector or Community-Supported Agriculture (CSA) in the suburbs of Beijing proves that China remains a decentralized country, where pockets of innovation and lagging exist. In the county of Huangmo, in Ningxia, poor environmental and economic conditions prevent the entrepreneurial model to emerge but at the same time allow other models to develop instead. These latest gather a wide variety of players belonging to state and nonstate circles, between which boundaries are often blur—which is also the case in the grain sector, traditionally managed by the state. In Beijing, rising food safety concerns of wealthy urban consumers encouraged the development of CSA farms, which continuously innovate in terms of sustainable farming practices, holding out the hope that alternative agriculture consuming less pesticides and chemical fertilizers and using less resources emerges. However, the existence of these fragmented pieces of territory does not put back into question the dominant frame of reference for agricultural modernization but rather demonstrates that the domination of a model for modernization is never incompatible with the existence of other models, which are almost unavoidable in a fragmented political and social environment such as China. In addition, the "alternative models" that were investigated were not fundamentally putting back into question the dominant frame of reference, as in Ningxia, local officials kept on referring to the other elements of the dominant frame and farmers were still locked in their social position, while Beijing innovative CSA farms were neither likely to spread their sustainable model to the whole country, nor fundamentally proposing alternative solutions to social issues brought by the dominant frame of reference for agricultural modernization.

7.4   PERSPECTIVES FOR THE FUTURE

Even though this research is a preliminary analysis that would need to be supplemented by quantitative data, it constitutes a solid base defining the social frames of agricultural modernization in China and shedding light on several theoretical frameworks in political science. In addition, this research gives a number of elements on the features of the pathway on which the Chinese agricultural sector is engaging, which is characterized by a certain number of rigidities that are likely to stand the test of time. According to the sayings of a number of central officials, the situation of deserted areas is unlikely to change in very drastic ways. Arbitration for financial efforts needs to be done. The regions characterized by the fragility of their ecosystem or by their low potential for agricultural development are unlikely to be better supported by the central government in the future. This hypothesis is further strengthened by the policy guidelines promoted in recent central documents. As it is indeed emphasized in the 12th Five-Year Plan, strategic regions for agricultural production should be given priority for modernization: "Optimize agricultural production and accelerate the building of the system in [...] agricultural main production areas [...] 'the seven areas and twenty-three zones [七区二十三带 *qi qu ershisan dai*]'".

Pockets of innovation, on their side, will keep on existing: firstly, because of the fragmentation of the Chinese state and the diversity of interests of local officials; and secondly, because experimentation is in itself a model for evolution in China, as explained by a large body of literature (Heilmann 2008a, b; Rawski 1995). However, for now, these areas of innovation and the models they developed poorly put back into question the economic and social marginalization of farmers and seem unable to scale up and to address environmental issues brought by agricultural modernization at the national level.

Nevertheless, questions related to the possible evolution of the dominant agricultural modernization model are worth asking. As we saw, the spreading of certain elements of the dominant frame of reference designed by the central government had consequences on the agricultural modernization pathway that are likely to endure in the middle and long term. The fact that the sector engaged on this pathway indeed created a structural inflexibility that now prevents the country from engaging on more sustainable trajectories. At the same time, the fissuring of the state's legitimacy is likely to make the central government increasingly in search of new sources for adaptation and power restoration. Until today, one of the most important sources for the legitimacy of the CCP was its ability to generate economic growth. Laliberté and Lanteigne (2008: 8–10), for instance, argue that economic performance is one of the three bases of the CCP's claims to legitimacy. Even the authors who consider that economic growth as the basis of the CCP's legitimacy is an oversimplification admit that the importance attached to economic growth has remained high among the population (Holbig and Gilley 2010). Even though economic development also took place in the countryside, rural areas are usually considered as having been left out of the process. A lot of scholars have emphasized that there was in reality a multiplicity of sources of legitimacy for the Chinese state apart from economic growth. Wang (2012), for instance, argues that state legitimacy could be enhanced by reforms of the state administration, such as the supervision and accountability system of cadres. "Disciplining officials" is also described by Tong and Lei (2014) as a common state response to social protests—one lever that was extensively used in the most recent years. A number of analyses also reach the conclusion that state legitimacy could be improved by the evolution of modes of governance (with some authors acknowledging that, although being an authoritarian state, China had proven able to develop forms of democracy and a pluralization of the political process in order to enhance its legitimacy (Guo 2010)) or by ideology-based arguments (Deng and Guo 2011). For Hibou (2011), state power is not only about obedience and prohibition imposed from above but can also aim at pleasing desires, bringing in "positive elements that influence the behavior of citizens." For her, the desire of modernization leads to the desire of state and therefore constitutes an important vector of domination. Traditionally, the notion of modernization is associated with the development of capitalism and with urbanization. In China, although agriculture was considered as one of the main pillars of economic development during the Maoist era, the 1980s and 1990s acknowledged the triumph of the classical view, according to which urban areas are the best representatives of modernity (Yeh et al. 2013: 917)—agriculture and the countryside turning into everything but places where modernization can be expressed. In the last decade however, we acknowledged a return of modernization discourses for rural areas. Through the development of private agricultural entrepreneurship led by private entrepreneurs and dragonhead enterprises progressively building a modern and industrialized food chain, a new wave of modernization reached rural areas and got peasants on the board of modern China, likely to strengthen the legitimacy of the state.

However, even though such levers for legitimacy exist and even though the Chinese state has proven a strong capacity to find new sources for legitimacy in the past (Laliberte and Lanteigne 2008; Guo 2010; Heberer and Schubert 2009; Tong and Lei 2014), at least two questions arise. The first one is linked to the possibility, for government officials, to link up with new circles of economic players—farmers—considering the importance of cultural factors, path dependencies, and local sets of interests. Not taking into account the point of view of small farmers could be particularly detrimental for the legitimacy of the Chinese state, not only in rural areas but in urban areas as well. The second question is about the practical capacity of central and local governments to maintain their level of financial support for agricultural development. Among others, questions arise about the financial capacity of the state to keep on supporting agricultural development. When I conducted fieldwork, debates were intense about the question of the abolition of minimum prices for grain. Even though these debates were not solely motivated by financial purposes—there were also market distortion issues—questions about the financial capacity of the Chinese state to keep on supporting agriculture on a sustainable way are worth raising, given the fact that the "state-led and export driven model has now almost exhausted its potential" (Yu Yongding, former director of the Institute of World Economics and Politics at CASS, cited in Li 2011).

Challenges therefore remain in the field of agriculture and state legitimacy. According to Almond and Powell (1966), there are five dimensions of state capacity: extractive, regulative, distributive, symbolic, and responsive. The Chinese state, on the side of agricultural modernization, still has to prove its ability to develop regulative, distributive, and responsive capacities. According to Remick (2004: 12), state-building is "the process in which state actors make a state organization grow in size, extend its reach and increase its functions". It is not sure yet whether the state-building process that started in rural areas with the 2004 Number One Document will last over time, as, for now, it still impedes the majority of small farmers from taking part in agricultural modernization.

What factors and what kind of change are likely to help the country overcome this challenge in the future? A wide body of literature on change can provide some answers to this question. In particular, this literature insists on the fact that the institutional, regulatory, and social context is not fixed and may vary over time and trigger change. For instance, the evolution of policy guidelines and the implementation of new political tools in the past proved that they could trigger change. New policy guidelines promoted by central documents and new financial support tools indeed had important effects on the modernization of the agricultural sector over the last decade. In the past, the Chinese government demonstrated a strong preference for gradual reforms (Liew 1995)—as opposed to "shock therapy". However, given the considerable challenge brought by environmental issues and their probable consequences on national food security in the middle term, major changes are likely to happen in the near future. Concrete reforms of the *hukou* scheme are already promulgated, considering the urgency to provide solutions to the unbearable situation of migrant farmers. It is not sure however whether access to social security and education will outweigh a guaranteed allocation of agricultural land. In addition, the Chinese government recently decided to promote family farming as another sociological tool to steer agricultural modernization, alongside dragonhead enterprises and farmers' associations. Family farming, which theoretically excludes industrial enterprises, could give a new importance to farmers and make them become real economic players of agricultural modernization. However, as the example of farmers' cooperatives depicted in this book demonstrates, change mainly comes from stakeholders and social structures. The fact that sets of interests, preferences, and strategies are never fixed and can vary according to contextual evolutions—for instance, the establishment of new regulations—holds out hope though that the situation evolves.

Change can also come from institutions, through administrative reforms. Even though the past evolution of the Chinese governmental structure left agricultural administrations relatively unimpaired, changes inside the general administration of the state could still affect the mode of implementation of agricultural policies. For instance, rising environmental issues increasingly push the central government to revise the cadres evaluation system, which could, in turn, lead to stricter supervision of agricultural policy implementation or to a rise in importance of environmental evaluation criteria—even though a number of studies proved that this mechanism had been inefficient so far (Wang 2013).

As we see, administrative reforms, the promulgation of new policy guidelines, the development of new political and sociological tools are likely to bump against sociological obstacles and local path dependencies. The implementation of change is often limited by the set of interests and the capacity to act of local officials and the effects of policies are likely to be narrowed down by the resilience of cultural factors (Murphy 2004; Anagnost 2004; Thogersen 2003; Kipnis 2006). As phrased by Bezes and Le Lidec (2010: 70): "Institutional reforms [...] often presented as drivers of change, [...] often do not necessarily affect power patterns, rules or games inherited from the past."[1] For the authors (2010: 86), the first necessary condition for the emergence of institutional reforms is the identification of "reform entrepreneurs",[2] capable of reconciling conflicting points of view and enabling compromise to be reached, in order to persuade large groups of actors to be part of a support coalition. Rural agri-food entrepreneurs could have played such a role of "reform entrepreneurs". However, the distance that is put between them and farmers-workers limits their ability to persuade these groups of actors to support the "curbing" of the current unsustainable agricultural pathway. A lot of work remains to be done to identify these stakeholders of change.

7.5   Going Further: Contributing to Global Debates

In the 1960s and 1970s, the promotion of the "Green Revolution" as a solution to address global hunger contributed to the emergence of input-intensive agricultural production models, of which the environmental and social limits have recently been widely denounced. According to Griffon (2002), not only has the Green Revolution been unable to reach poor people—particularly landless peasants—in Asia and in South America but it also caused a disastrous environmental degradation that now threatens the possibilities to ever achieve the first objective of the Green Revolution: solving global hunger. The food price crisis of 2007–2008 revived the debates by proving that agricultural and food security issues are still to be addressed, both in developing and in developed countries. Since then, agricultural pathways arouse considerable controversy around the world. Although none of the agricultural transition models discussed in international debates can be silver bullets, defining models and debating about their features are essential, because it gives a vision for the evolution of the agricultural sector—an evolution that has now become necessary.

By putting forward the sociological and cultural obstacles impeding the evolution of the agricultural sector toward more sustainable practices, this research wishes to inform the debate on the necessity to take into account stakeholders and to use sociology and behavioral studies. It also wishes to warn against the risk to believe that implementation of standard political, economic, and technological reforms will necessarily trigger change. As such, the 2015 World Development Report "Mind, Society and Behavior", which emphasizes that "development policies based on new insights into how people actually think and make decisions will help governments and civil society achieve development goals more effectively" (World Bank 2015) is a major step forward.

In addition, by showing the considerable importance of involving small farmers in agricultural transition and by underlining the difficulties agri-food enterprises are experiencing to become real "reform entrepreneurs", this research intends to contribute to the foundation of a corpus of research stressing the need to give a role to small farmers in agricultural transitions. Getting small farmers on board is not only a way to ensure the effective implementation of more sustainable farming practices. It is also a mean to enrich the general knowledge on sustainable farming practices. As under-lined by De Schutter (2010: 18): "Rather than treating smallholder farm-ers as beneficiaries of aid, they should be seen as experts with knowledge that is complementary to formalized expertise". At the beginning of the twentieth century, King (1949) was already giving tribute to the richness of local practices—his book, *Farmers of Forty Century*, was first published in 1911. Today, more than ever, the value of these practices needs to be better acknowledged.

The analysis of the Chinese case demonstrates that frames of reference promoted by agricultural policies and local patterns of power are likely to hinder the participation of small farmers in agricultural transformation. Even when there is willingness of central governments to implement solu-tions to voice out the views of farmers, it is often not sufficient to trigger change. Could it be possible to voice out their views and to frame the col-lective action of small farmers, for instance, inside international forums that could influence, at the same time, the action of national governments and the one of local actors? Which tools would allow to better communicate with marginalized groups of small farmers? These questions, which are clearly not unique to China, would deserve another book.

NOTES

1. Original language: "Les réformes institutionnelles, [...] souvent présentées au plan rhétorique comme le moteur de profonds changements, [...] n'ont pourtant pas nécessairement pour effet de modifier les structures de pouvoir, règles ou jeux antérieurs."
2. Original language: "Entrepreneurs de réforme."

REFERENCES

Almond, G. A., & Powell, G. B. (1966). *Comparative politics: A developmental approach*. Boston: Little, Brown.

Anagnost, A. (2004). The corporeal politics of quality (suzhi). *Public Culture, 16*(2), 189–208. https://doi.org/10.1215/08992363-16-2-189.

Bezes, P., & Le Lidec, P. (2010). Ce que les réformes font aux institutions. In J. Lagroye & M. Offerle (Eds.), *Sociologie de l'institution*. Paris: Belin.

De Schutter, O. (2010). Report submitted by the Special Rapporteur on the right to food, A/HRV/16/49, Geneva.

Deng, Z., & Guo, S. (2011). *Reviving legitimacy: Lessons for and from China*. Lanham/Boulder/New York: Lexington Books.

Griffon, M. (2002). Révolution Verte, Révolution Doublement Verte: Quelles technologies, institutions et recherche pour les agricultures de l'avenir ? *Mondes en développement, 1*(117), 39–44.

Guo, B. (2010). *China's quest for political legitimacy: The new equity-enhancing politics*. Lanham: Lexington Books.

Heberer, T., & Schubert, G. (Eds.). (2009). *Regime legitimacy in contemporary China: Institutional change and stability*. New York/London: Routledge.

Heilmann, S. (2008a). Policy experimentation in China's economic rise. *Studies in Comparative International Development, 43*(1), 1–26.

Heilmann, S. (2008b). From local experiments to national policy: The origins of China's distinctive policy process. *The China Journal, 59*, 1–30.

Hibou, B. (2011). *Anatomie politique de la domination*. Paris: la Découverte.

Holbig, H., Gilley, B. (2010). *In search of legitimacy in post-revolutionary China: Bringing ideology and governance back in*. GIGA working papers 127.

King, F. (1949). *Farmers of forty centuries; or, permanent agriculture in China, Korea and Japan*. London: J. Cape.

Kipnis, A. (2006). Suzhi: A keyword approach. *The China Quarterly, 186*, 295–313.

Laliberte, A., & Lanteigne, M. (Eds.). (2008). *The Chinese party-state in the 21st century: Adaptation and the reinvention of legitimacy*. London/New York: Routledge.

Landry, P. F. (2008). *Decentralized authoritarianism in China: The Communist Party's control of local elites in the post-Mao era*. Cambridge/New York: Cambridge University press.

Li, C. (2011). Introduction: A champion for Chinese optimism and exceptionalism. In A. Hu (Ed.), *China in 2020: A new type of superpower*. Washington, DC: Brookings Institution Press.

Liew, L. H. (1995). Gradualism in China's economic reform and the role for a strong central state. *Journal of Economic Issues, 29*(3), 883–895.

Murphy, R. (2004). Turning peasants into modern Chinese citizens: 'Population quality' discourse, demographic transition and primary education. *The China Quarterly, 177*, 1–20.

Rawski, T. G. (1995). Implications of China's reform experience. *The China Quarterly, 144*, 1150–1173.

Remick, R. J. (2004). *Building local states: China during the republican and post-Mao eras*. Cambridge, MA/London: Harvard University Asia Center.

Thogersen, S. (2003). Parasites or civilizers: The legitimacy of the Chinese Communist Party in rural areas. *China: An International Journal, 1*(2), 220–223. https://doi.org/10.1353/chn.2005.0038.

Tong, Y., & Lei, S. (2014). *Social protest in contemporary China, 2003–2010: Transitional pains and regime legitimacy*. Abingdon/Oxon/New York: Routledge.

Wang, P. (2012, November 11). Breaking and building rural finance. *Caijing Magazine*. [王培成, "农村金融破与立", 《财经》杂志, 11/11/2012. *Wang Peicheng, Nongcun jinrong po yu yi, Caijing Zazhi*].

Wang, A. L. (2013). The search for sustainable legitimacy: Environmental law and bureaucracy in China. *The Harvard Environmental Law Review, 37*(2), 365–440.

World Bank. (2015). *World development report: Mind, society and behavior*. Washington, DC: World Bank.

Yeh, E. T., O'Brien, K. J., & Ye, J. (2013). Rural politics in contemporary China. *The Journal of Peasant Studies, 40*(6), 915–928. https://doi.org/10.1080/03 066150.2013.866097.

Appendix: Official Documents

Chinese Official Documents

Five-Year Plans

Seventh Five-Year Plan for National Economic and Social Development of the Republic of China (1986–1990) [中华人民共和国国民经济和社会发展第七个五年计划 *zhonghua renmin gongheguo guomin jingji he shehui fazhan di qi ge wunian jihua*].

Eighth Five-Year Plan for National Economic and Social Development of the Republic of China (1991–1995) [中华人民共和国国民经济和社会发展十年规划和第八个五年计划 *zhonghua renmin gongheguo guomin jingji he shehui fazhan di ba ge wunian jihua*].

Ninth Five-Year Plan for National Economic and Social Development of the Republic of China (1996–2000) [中华人民共和国国民经济和社会发展十年规划和第九个五年计划 *zhonghua renmin gongheguo guomin jingji he shehui fazhan di jiu ge wunian jihua*].

Tenth Five-Year Plan for National Economic and Social Development of the Republic of China (2001–2005) [中华人民共和国国民经济和社会发展十年规划和第十个五年计划 *zhonghua renmin gongheguo guomin jingji he shehui fazhan di shi ge wunian jihua*].

© The Author(s) 2018   251
M.-H. Schwoob, *Food Security and the Modernisation Pathway in China*, Critical Studies of the Asia-Pacific,
https://doi.org/10.1007/978-3-319-65702-8

Eleventh Five-Year Plan for National Economic and Social Development of the Republic of China (2006–2010) [中华人民共和国国民经济和社会发展十年规划和第十一个五年规划 *zhonghua renmin gongheguo guomin jingji he shehui fazhan di shiyi ge wunian guihua*].

Twelfth Five-Year Plan for National Economic and Social Development of the Republic of China (2011–2015) [中华人民共和国国民经济和社会发展十年规划和第十二个五年规划 *zhonghua renmin gongheguo guomin jingji he shehui fazhan di shier ge wunian guihua*].

Number One Documents

2004 Number One Document—State Council's opinion about policies to accelerate the rise in farmers' income [2004年中央一号文件 – 国务院关于促进农民增加收入若干政策的意见 *2004 nian zhongyang yihao wenjian—guowuyuan guanyu cujin nongmin zengjia shouru ruogan zhengce de yijian*].

2005 Number One Document—State Council's opinion about policies to go a step further in the strengthening of rural work to raise the production capacity of agriculture [2005年中央一号文件 – 国务院关于进一步加强农村工作提高农业综合生产能力若干政策的意见 *2005 nian zhongyang yihao wenjian – guowuyuan guanyu jinyibu jiaqiang nongcun gongzuo tigao nongye zonghe shengchan nengli ruogan zhengce de yijian*].

2006 Number One Document—State Council's opinion on how to promote the building of the new socialist countryside [2006年中央一号文件—国务院关于推进社会主义新农村建设的若干意见 *2006 nian zhongyang yihao wenjian—guowuyuan guanyu tuijin shehui zhuyi xin nongcun jianshe de ruogan yijian*].

2007 Number One Document—State Council's opinion about how to actively develop modern agriculture and to promote the building of the new socialist countryside [2007年中央一号文件—国务院关于积极发展现代农业扎实推进社会主义新农村建设的若干意见 *2007 nian zhongyang yihao wenjian—guowuyuan guanyu jiji fazhan xiandai nonghe zhashi tuijin shehui zhuyi xin nongcun jianshe de ruogan yijian*].

2008 Number One Document—State Council's opinion about how to realistically reinforce the building of the basis of agriculture and go a step further in agricultural development and in the increase of farmers' income [2008年中央一号文件—关于切实加强农业基础建设进一步促进农业发展农民增收的若干意见 *2008 nian zhongyang yihao wenjian—guowuyuan guanyu qieshi jiaqiang nongye jichu jianshe jinyibu cujin nongye fazhan nongmin zengshou de ruogan yijian*].

2009 Number One Document—State Council's opinion about how to accelerate the stable development of agriculture and how to keep on increasing farmers' income [2009年中央一号文件—国务院关于2009年促进农业稳定发展农民持续增收的若干意 *2009 nian zhongyang yihao wenjian—guowuyuan guanyu 2009 nian cujin nongye wending fazhan nongmin chixu zengshou de ruogan yijian*].

2010 Number One Document—State Council's opinion about how to enlarge the dynamic of comprehensive rural and urban development and go a step further in the basis of agricultural and rural development [2010年中央一号文件—国务院关于加大统筹城乡发展力度进一步夯实农业农村发展基础的若干意见*2010 nian zhongyang yihao wenjian—guowuyuan guanyu jiada tongchou chengxiang fazhan lidu jinyibu hangshi nongye nongcun fazhan jichu de ruogan yijian*].

2012 Number One Document—State Council's opinion about how to accelerate the promotion of agricultural science and technology innovation and to keep on strengthening the ability to protect food security [2012年中央一号文件—国务院关于加快推进农业科技创新持续增强农产品供给保障能力的若干意见*2012 nian zhongyang yihao wenjian—guowuyuan guanyu jiakuai tuijin nongye keji chuangxin chixu zengqiang nongchanpin gongji baozhang nengli de ruogan yijian*].

2013 Number One Document—State Council's opinion about how to accelerate the development of modern agriculture and go a step further in the strengthening of the vitality of rural development [2013年中央一号文件—国务院关于加快发展现代农业　进一步增强农村发展活力的若干意见*2013 nian zhongyang yihao wenjian—guowuyuan guanyu jiakuai fazhan xiandai nongye jinyibu zengqiang nongcun fazhan huoli de ruogan yijian*].

2014 Number One Document—State Council's opinion about how to comprehensively deepen rural reform and accelerate the push for agricultural modernization [2014年中央一号文件—国务院关于全面深化农村改革加快推进农业现代化的若干意见*2014 nian zhongyang yihao wenjian—guowuyuan guanyu quanmian shenhua nongcun gaige jiakuai tuijin nongye xiandaihua de ruogan yijian*].

Other Central-Level Documents
Central Committee of the Communist Party of China (2014) Communiqué of the Third Plenary Session of the 18th Central Committee of the CPC. Beijing, 2014 http://www.china.org.cn/chinese/2014-01/16/content_31213800_3.htm

Chinese Ministry of Agriculture (Department of Sectoral Policy and Law) (2013) Policy Measures to Support Increase in Grain Output and Farmers' Income (Part II). Online newsfeed of the Ministry of Agriculture, 19 April 2013 http://english.agri.gov.cn/governmentaffairs/pi/201304/t20130422_19488.htm

Chinese State Council (2001) Outline for China Rural Poverty Alleviation and Development 2001–2010 [国务院关于印发中国农村扶贫开发纲要(2001–2010年)的通知 *Guowuyuan guanyu yinfa zhongguo nongcun fupin kaifa gangyao (2001–2010 nian) de tongzhi*] http://www.gov.cn/gongbao/content/2001/content_60922.htm

Chinese State Council (Research Office) (2006) *Research Report on Migrant Workers.* 2006 [国务院研究课题室, 中国农民工调研报告, 北京: 言实出版社, 2006 *Guowuyuan yanjiu keti shi, zhongguo nongmingong diaoyan baogao, beijing : yanshi chubanshe*].

General Administration of Quality Supervision, Inspection and Quarantine. *Of the ways of managing food production licenses (General Order n°129)* [国家质量监督检验检疫总局 "食品生产许可管理办法"(总局令第129号) *guojia zhiliang jiandu jianyan jianyi zongju "shipin shengchan xuke guanli banfa" (zongju ling di 129 hao)*] http://www.aqsiq.gov.cn/xxgk_13386/jlgg_12538/zjl/20092010/201210/t20121016_239328.htm

Ministry of Agriculture (2013) GM and non-GM food are similarly safe. News Office of the Ministry of Agriculture, 31 August 2013 ["转基因食品与非转基因食品具有同样的安全性," 农业部新闻办公室] http://www.moa.gov.cn/zwllm/zwdt/201308/t20130831_3592472.htm

Ministry of Agriculture (Information Office) (2013) Vice Minister Zhang prioritizes three major projects in agricultural sci-tech. Online newsfeed of the Ministry of Agriculture, 4 March 2013 http://english.agri.gov.cn/news/dqnf/201304/t20130409_12148.htm

Ministry of Agriculture (Department of Sectoral Policy and Law) (2013) Policy Measures to Support Increase in Grain Output and Farmers' Income (Part II). Online newsfeed of the Ministry of Agriculture, 19 April 2013 http://english.agri.gov.cn/governmentaffairs/pi/201304/t20130422_19488.htm

Ministry of Finance (2014) Financial support situation for the three rural issues. Beijing, 2014. [财政部,财政支持"三农"情况 *Caizhengbu, caizheng zhichi "sannong" qingkuang*] http://www.mof.gov.cn/zhuantihuigu/czjbqk1/czzc/201405/t20140507_1076149.html

Ministry of Land and Resources (2014a) Ministry of Land and Resources rectifies and reforms land rights violations in six local governments, published on the website of the Ministry of Land and Resources of China, 10 April 2014 [国土部限期六地方政府整改土地违规, 中国国土资源部网, 10/04/2014 *Guotubu xianqi liu defang zhengfu zhenggai tudi weigui*] http://www.mlr.gov.cn/xwdt/mtsy/qtmt/201404/t20140410_1311884.htm

Ministry of Land and Resources (2014b) Ministry of Land and Resources hangs out the shingle of land right violation of the county of Yongnian in Hebei, published on the website of the Ministry of Land and Resources of China, 16 May 2014 http://www.mlr.gov.cn/xwdt/mtsy/people/201405/t20140516_1317279.htm [国土部挂牌督办河北省永年县违法征占土地案, 中国国土资源部网, 16/05/2014 *Guotubu guapai duban hebeisheng yongnianxian weifa zhengzhan tudi an*].

People's Republic of China (2006) Law on Farmers' Specialized Cooperatives [中华人民共和国农民专业合作社法*Zhonghua Renmin Gongheguo Nongmin Huanye Hezuoshe Fa*] http://www.gov.cn/jrzg/2006-10/31/content_429182.htm

Report on 2008–2013 central and local budget situation and 2009–2014 draft for central and local budget [关于2008–2013年中央和地方预算执行情况与2009–2014年中央和地方预算草案的报告 *guanyu 2008–2013 nian zhongyang he difang yusuan zhixing hang qingkuang yu 2009–2014 nian zhongyang he difang yusuan cao'an de baogao*].

Local-Level Documents
2013 Name list of brokers of agricultural machinery for the subsidy system for the purchase of agricultural machinery of Zhanggong qu [2013年章贡区具备农机购置补贴产品经销资质的农机经销商名单 (2013 *nian zhanggong qu jubei nongji gouzhi butie chanping jingxiao zhidi de nongji jingxiaoshang mingdan*)].

["Huangmo"] Government (2012) Annual Agricultural Machinery Purchase Subsidy Process, published on the website of "Huangmo" government, 12 Dec 2012 ["Huangmo"] 2012 年农业机械购置补贴办理流程 *["Huangmo"] 2012 nian nongye jixie gouzhi butie banli liucheng*.

["Lanshui"] Office of Agricultural Machinery (2013) Subsidy policy for agricultural machinery purchase in 2013, published on the website of "Lanshui" Government, 24 June [2013 年第一批农业机械购置补贴政策简介 *["Lanshui"] 2013 di yi pi nongye jixie gouzhi butie zhengce jianjie*].

["Lushan"] Bureau of agricultural machinery (2013) Municipal agricultural bureau reforms the procedure for subsidies for the purchase of agricultural machinery, published on the website of ["Lushan"] government, 1 April 2013 [市农机局改革农机购置补贴程序 *shi nongjiju gaige nongji gouzhi butie chengxu*].

["Lushan"] Bureau of farm machinery (2014) Establishing a three items-management system for policies regarding subsidies for farm machinery purchase, published on the website of ["Lushan"] government, 24 Feb 2014 [农机局:制定农机购置补贴政策实施三项管理制度 *[Lushan] nongjiju : zhiding nongji gouzhi butie zhengce shishi sanxiang guanli zhidu*].

["Lushan"] Government (2014a) Organizing trainings for subsidies for the purchase of farm machinery, published on the website of ["Lushan"] Government, 12 February 2014 [举办全市农机购置补贴培训班 *juban quanshi nongji gouzhi butie peixun ban*].

["Lushan"] Government (2014b) ["Lushan"] simplifies the procedure for subsidies for farm machinery purchase, published on the website of ["Lushan"] government, 13 Feb 2014 [简化农机购置补贴发放程序 *[Lushan] jianhua nongji gouzhi butie fafang chengxu*].

Ningxia Government (2012) The 2012 GDP of the autonomous region reaches 240 billion yuan, Government of Ningxia, 16 Dec 2012 [2012年自治区GDP预计达到两千四百亿元，　宁夏自治区人民政府, 26/12/2012 *2012 nian zizhiqu GDP yuji dadao liang qian si bai yi yuan, ningxia zizhiqu renmin zhengfu*] http://www.nx.gov.cn/zwxx/zw/zwdt/47370.htm

Ningxia procedure for subsidies for the purchase of agricultural machinery [宁夏农业机械购置补贴办理流程 *ningxia nonye jixie gouzhi butie banli liucheng*].

Yantai Government (2012) Information on Yantai's implementation program of subsidies for the purchase of agricultural machinery [关于印发2013年烟台市农业机械购置补贴工作实施方案的通知　*guanyu yinfa 2013 nian yantaishi nongye jixie gouzhi butie gongzuo shishi fang'an de tongzhi*].

Index[1]

[1] Note: Page numbers followed by "n" refer to notes.

Printed by Printforce, the Netherlands